Managing Intellectual Assets
in the Digital Age

For a listing of recent titles in the *Artech House Communications Law and Policy Series*, turn to the back of this book.

Managing Intellectual Assets in the Digital Age

Jeffrey H. Matsuura

Artech House
Boston • London
www.artechhouse.com

Library of Congress Cataloging-in-Publication Data
Matsuura, Jeffrey H., 1957–
    Managing intellectual assets in the digital age / Jeffrey Matsuura.
        p. cm.—(Artech House communications law and policy series)
    Includes bibliographical references and index.
    ISBN 1-58053-359-0 (alk. paper)
        1. Intellectual property. 2. Copyright and electronic data processing.
    3. Copyright—Computer programs. 4. Software protection—Law and legislation.
    I. Title. II. Artech House communications law and policy library.

K1401.M378 2003
346.04'8—dc21   2003041495

British Library Cataloguing in Publication Data
Matsuura, Jeffrey H., 1957–
    Managing intellectual assets in the digital age.—(Artech House communications
    law and policy series)
    1. Intellectual property 2. Intellectual property—Law and legislation
    I. Title
    346' .048

    ISBN 1-58053-359-0

Cover design by Igor Valdman

© 2003 ARTECH HOUSE, INC.
685 Canton Street
Norwood, MA 02062

International Standard Book Number: 1-58053-359-0
Library of Congress Catalog Card Number: 2003041495

10 9 8 7 6 5 4 3 2 1

With all my thanks for their boundless support and enthusiasm, this book is dedicated to Harry and Tomeko Matsuura, and to Janice, Anne, Fuzz, and Beau—a team that makes the extraordinary possible.

Contents

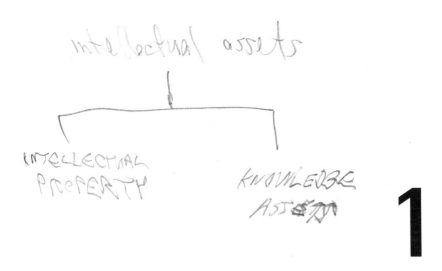

1

Introduction: Overview of intellectual property and knowledge assets

1.1 Purpose of this book

This book examines challenges and opportunities associated with the development, distribution, and use of intellectual property and knowledge assets. For purposes of this book, intellectual property consists of the creative works that are governed by the principles of traditional intellectual property law: copyrights, patents, and trademarks. Knowledge assets consist of information, knowledge, and know-how that have commercial value but that have not traditionally been governed by intellectual property law principles of copyrights and patents. The intellectual assets that are the focus of this book include both traditional intellectual property and knowledge assets.

Strategies to manage intellectual assets are in a state of flux, largely as a result of the expansion of the Internet and associated computer networks. Wide-scale computer networking alters the process of creation, distribution, modification, and use of intellectual assets. Networking changes the supply of, and demand for, many intellectual assets. In this way, digitization of intellectual assets and integration of those assets into computer networks affect the economic value of the assets. Networks also affect the ability of developers and distributors of intellectual assets to control user access to those assets. Management of access to those assets (digital rights management) is a major challenge in the networked environment.

This book examines the most significant digital rights management challenges triggered by global computer networking, and it identifies some of the developing rights management strategies aimed at addressing those challenges. Those management strategies begin with traditional intellectual property law principles but also make use of other legal theories and economic incentives. Basic intellectual property law forms the foundation for digital rights management, supplemented by a variety of other legal concepts and commercial strategies. This book provides developers, distributors, and consumers of intellectual assets with information regarding their rights and duties as to the control and use of those assets.

1.2 Who should read this book, and why?

This book can be most helpful for those who have a professional or personal interest in the creation, distribution, or consumption of intellectual assets. That interest may arise from a wide range of sources. It can, for example, stem from a direct interest (e.g., readers who are actual creators or users of intellectual assets, whether they are software developers, Web content creators, or online music aficionados). The interest can also arise from an indirect interest in intellectual assets (e.g., readers who are associated with "intermediaries" or are involved in the distribution of intellectual assets, such as online service providers or media enterprises). Other people who may have an indirect interest in management of intellectual assets are the people who finance the creation or use of those assets (e.g., investors, lenders). Finally, another important group of interested parties includes those involved in the development or enforcement of public policies that affect creation or use of intellectual assets (e.g., legislators, judges, and regulators).

As to the question of why read this book, the answer is relatively simple. Read this book so that you will better understand what intellectual assets are, what current and developing challenges (both legal and economic) affect their creation and use, and what effective strategies for managing commercial use of those assets are emerging. This book does not provide detailed analysis of any one issue but instead attempts to present an understandable overview of many of the most important intellectual asset management issues being faced today. It is not a treatise but is instead a survey, a primer, or a guide. The book does not offer legal advice, for that the reader should consult counsel familiar with the specific needs and circumstances of the reader. Instead, the book is intended to provide information that can be quickly digested by the reader and promptly applied to the immediate needs of the reader. This book will not give you the answers to all of your intellectual asset management issues; however, it will help you to understand what rights, responsibilities, and opportunities you should be aware of as you develop your strategies for development, management, and use of those assets.

1.3 Intellectual property law principles

An important aspect of this book is the relationship between well-established principles of intellectual property law and the management of intellectual assets. Traditional forms of intellectual property have long been protected under copyright, patent, and trademark laws. In order to understand rights and duties associated with control of intellectual assets, a basic understanding of fundamental principles of intellectual property law is required. The law of intellectual property balances the interests of the creators of intellectual property with those of the users of that property. The law provides developers of intellectual content with the exclusive right to exploit that content commercially. This grant of ownership is intended to provide creators of intellectual property with continuing economic incentive for future creative work. Intellectual property law also established certain limited rights of use (e.g., fair use) vested in the users of that property, which could not be eliminated by the creators of that property. This grant of rights to users of intellectual property was intended to promote rapid integration of creative works into new products, resulting in economic growth. In this book, we provide an overview of the ways in which traditional intellectual property law principles are being applied as part of digital rights management initiatives. That discussion examines the impact of application of traditional intellectual property law rights on digital content developers, distributors, and users.

Copyright law provides rights of ownership to the authors of original works that have been fixed in tangible form. Copyright owners possess the following rights with respect to their works: the right to duplicate the work; the right to distribute the work;the right to display, exhibit, or perform the work; and the right to create "derivative works" (i.e., works that are based on the original work, such as a motion picture version of a book). All of these basic rights are held by the author of the original work immediately upon creation of the work. Among the three core forms of intellectual property law rights, copyright law has been, to date, the most actively applied in the context of digital rights management. As a result, digital copyright issues form an important component of this book.

Copyright law also provides important basic rights for users of copyright protected works. The concept of "fair use" permits individuals to make use of copyrighted works, without the permission of the author, under certain limited circumstances. For example, the doctrine of fair use permits individuals to use portions of the works for educational or scholarly purposes without permission. Fair use also permits use of excerpts of copyrighted works for commentary, news reporting, or satire. All fair use of copyrighted material must be noncommercial in nature and must make only limited use of the material (i.e., use only short excerpts of the material, not the entire work, and limit distribution of the material). Efforts to apply traditional protections for users established by copyright law, including fair use, will be an important and controversial element of future digital rights management debates. Key aspects of this debate are addressed in this book.

Patent law provides rights of ownership to the creators of new inventions that are novel, useful, and nonobvious. Traditionally, patent law provided rights for the inventors of machines (e.g., equipment), processes (e.g., manufacturing procedures), and materials. Patent law also provided rights to the creators of novel designs for products (e.g., design patents). More recently, patent rights have been granted to an expanding array of inventions (e.g., computer programs, business methods, modified living organisms). Some of these newer patent fields have created controversy. For example, business method patents, the patents that cover business practices and processes (e.g., fund allocations, payment systems, billing processes), have inspired a substantial amount of discussion and debate.

Unlike copyrights, patent rights do not exist until an inventor has applied to the government for those rights, and the government has granted the patent rights based on a determination that the invention satisfies the basic patent requirements. After obtaining patent rights, the inventor has the exclusive right to "practice" the invention (i.e., to manufacture, distribute, and use the invention). This right enables the owner of the patent to block other parties from practicing an invention that falls within the scope of the patent. The scope of an issued patent covers the description of how the invention is made and how it functions, along with the "claims" incorporated into the patent describing what functions the invention performs.

In the past, patent rights came into play in the context of computers and their content, primarily with regard to the development and manufacture of computer equipment. That trend is rapidly changing. Although patent rights applied to computer hardware continue to be a critically important subject, patent rights are also increasingly important in the context of computer software and electronic commerce systems. For example, a list of the most controversial patent law issues associated with the digital rights environment would include the challenge of defining the appropriate scope of patent rights applicable to computer programs and to business methods in the e-commerce setting. Patent law has thus established a much broader scope of applicability with regard to digital rights management than it has traditionally held, and it is reasonable to expect its expanding reach to become still more diverse in the future. These emerging digital patent law issues are addressed in this book.

A third traditional form of intellectual property rights is trademark law. Trademark law permits the developers of commercial names, slogans, logos, and a wide range of other marks to assert enforceable ownership rights over those marks. Trademark law protects commercial marks from use by other parties that could result in confusion on the part of the public. Trademarks are viewed to be commercial identifiers (or signatures) that enable the consuming public to recognize specific companies and their products. Trademark law protects the economic investment that businesses make in creating their commercial marks by empowering those businesses to block use of similar trademarks, when such other use could lead to consumer confusion.

Trademark law also includes the concepts of trade dress and trademark dilution. Trade dress is the portion of trademark law that governs the appearance of packaging and product design. Trademark dilution law permits trademark owners to block use of identical or similar marks when those other uses dilute the economic value of the original mark. Dilution rights can be applied even in those cases where the other use of the mark is not likely to result in consumer confusion. Trademark dilution laws expand the scope of trademark ownership rights, enhancing them so that they protect the economic value of the mark itself, not just the effectiveness of the mark as a commercial identifier.

To date, trademark rights in the digital environment have been enforced primarily with regard to Internet domain names. Trademark laws, supplemented in some jurisdictions by regulations prohibiting domain name "cybersquatting" have been actively applied to help trademark owners protect their marks from misuse in domain names. In the future, however, we will likely see greater use of trademark law (including dilution and trade dress aspects of trademark law) to assist in the management of an expanding array of digital content. For example, we already see trademark law concepts applied to online keywords and the on-screen appearance of user interfaces. Expect this trend toward broader application of trademark law rights for digital media content to continue in the future. We will examine important aspects of the interplay between trademark law and intellectual assets in this book.

Digital rights management issues and strategies are profoundly affected by these three fundamental forms of intellectual property law. In this book, we examine some of the most important examples of how copyright, patent, and trademark law principles are currently influencing digital rights management. The book also explores the ways in which the expanding universe of digital content will likely test and alter long-held principles of intellectual property law rights in the future. An essential element of any future strategy for management of digital rights is a basic understanding of the principles of intellectual property law rights, and an appreciation of how those rights are dramatically influenced by the expansion of digital content. This book addresses those important elements.

1.4 Knowledge assets

We now realize that there are many forms of knowledge and information that have significant economic value but are not governed by traditional intellectual property laws. Operational know-how and collections of information, for example, can have substantial commercial value; however, they are not readily protected under traditional intellectual property law principles. Instead of protection under copyright, patent, or trademark laws, many of these knowledge assets are managed using legal principles, including trade secrets, personal property rights, commercial transactions law, antitrust/competition law, and recently established database ownership regulations. In this book, we examine the ways in which these legal concepts are now applied to digital rights management strategies and to broader strategies associated with knowledge management.

The law of trade secrets has long been applied to provide legal rights of ownership and control over information or knowledge that provides a commercial advantage. Trade secrets law provides remedies for parties who suffer commercial harm as a result of unauthorized disclosure of proprietary commercial material. The law of trade secrets is highly active in the context of digital content. In this book, we examine how regulation of trade secrets provides an effective tool for the developers of valuable information in the networked environment.

There is a growing trend toward application of traditional concepts of personal property law to intellectual assets. Legal theories such as "trespass" are now commonly applied as remedies for unauthorized access to computers and their content. Property law is developing a broader role in the process of intellectual asset management, and that expanding role is likely to continue into the future. We discuss, in this book, some of the different contexts in which property law principles are being applied to the management of intellectual assets.

The law of contracts and commercial transactions also plays an expanding role in the management of intellectual assets. As intellectual assets become the subject of a rapidly increasing number of commercial sales, contract law and the law of commercial sales have been called upon to govern the relationships between the buyers and sellers of intellectual assets. In the United States, for example, the effort to adopt a standard law for commercial sales of digital content at the state level [i.e., the Uniform Computer Information Transactions Act (UCITA)] has resulted in significant controversy and debate. It appears that this debate is only beginning and remains far from resolution. This book includes discussion of the developing role of commercial law in the management of intellectual asset transfers.

We also see that the law of antitrust and commercial competition plays a growing role in intellectual asset development and management. Antitrust and competition laws provide a legal framework to protect fair competition in the commercial marketplace. As intellectual assets have become more obviously important in commerce, antitrust and competition laws have been called upon, more frequently, to resolve controversies associated with development and distribution of intellectual assets. This book reviews some of the different ways in which the laws designed to protect commercial competition affect intellectual asset management strategies.

The law now also provides a broader scope of coverage for databases and other collections of information. Led by the European Community's Database Directive, there is now a clearer set of legal rights associated with the development and ownership of aggregations of information that have commercial value. In this book, we examine some of the ways in which those legal rights affect access to, and use of, information. We also review some of the more important open issues associated with database rights that remain to be resolved.

1.5 Emerging digital rights management issues

There are already several highly visible examples of digital rights management controversies that have been underscored by global networking. Those controversies include software patents, domain name conflicts, expansion of "Open Source" software development models, business method patents applied to electronic commerce, online digital media distribution, database ownership rights, and broader use of "peer-to-peer" content sharing in distributed computing environments. This book examines these, and other, examples of the increasingly complex digital rights management issues that are currently emerging. As we discuss those emerging issues, we will examine the legal rights and obligations that affect the resolution of those challenges. We will also review some of the important economic concerns raised by these digital rights issues.

It is also becoming increasingly apparent that digital rights management issues have significant international components. Globalization of network access and use make the creation, distribution, and consumption of intellectual assets international activities. In that setting, digital rights management strategies are profoundly influenced by legal and economic forces present in many different countries. This book examines some of the most important international aspects of the opportunities and challenges posed for intellectual assets by the development of global computer networks. Discussion of the internationalization of intellectual asset management strategies is an important element of this book.

2

Traditional software protection strategies

Software developers have traditionally relied on three forms of legal protection for their products: copyrights, patents, and trade secrets. Each of those three legal principles has advantages and disadvantages from the perspective of the developer of the software. This chapter examines the relative merits of the three approaches to software protection and their impact on both creators and users of computer programs.

2.1 Copyrights and software

Copyright law provides ownership and control to the creators of original works that are fixed in tangible form. It grants those creators several fundamental rights of use for the works they create. One of those rights is the right to create copies of (duplicate) the work. Another of those rights is the right to distribute the work. Copyright law also grants the creator of an original work the right to perform or to exhibit the work publicly. Finally, copyright law gives the creator of the work the right to create "derivative works" (i.e., works that are based on the original copyrighted work—a foreign language translation of a novel, for instance).

Copyright law protects only original works of authorship. It does not protect works that are not original or works that do not exhibit the creative element required for a determination of authorship. For instance, mere collections of facts or information are not governed by copyright law. In the United States, the leading court case that established the interpretation that mere collections of information were not subject to copyright law protections was *Feist Publications, Inc. v. Rural Telephone Service Company*. In that case, the U.S. Supreme Court concluded that a listing of names and telephone numbers in a directory did not qualify as a work of authorship eligible for copyright protection. Even though substantial resources may be required to create and maintain such compilations of information, copyright law does not protect those works. Other legal principles, however, can now be applied to such collections, as discussed more fully in Chapter 8.

One who exercises any of the rights associated with copyright without the permission of the copyright owner is an "infringer." The party directly responsible for the infringement is guilty of "direct" infringement. Parties who facilitate the direct infringement may also be guilty of copyright infringement. For example, a party who has the ability to supervise or control the conduct of the direct infringer can be liable for "vicarious" copyright liability based on the conduct of the direct infringer. If, for instance, an employee of a business engages in conduct that constitutes copyright infringement, the employer could be liable for copyright infringement based on the vicarious liability concept. In addition, a party who in some way facilitates or contributes to the distribution of infringing material can be held liable for copyright infringement under the principle of "contributory" infringement. Based on the theory of contributory infringement a book publisher, for example, could be held liable for infringing material contained in a book written by another party that it published.

Copyright law permits limited use of copyright protected material without the prior permission of the copyright owner if the use of the material qualifies as "fair use." Essentially, fair use consists of noncommercial use for educational, scholarly research, news reporting, commentary, or satirical purposes. Even if the use of the material qualifies as fair use, however, the user does not have unrestricted rights of use. The user claiming fair use rights must make sure that the use of the material is noncommercial, uses as little of the copyrighted material as possible, and does not adversely affect the commercial market for, or the economic value of, the copyrighted work.

Copyright law is applied to protect creative works in all of the countries that have major economic power. It protects the authors of original works and grants them exclusive rights of exploitation for the material that they create. International copyright ownership is governed by the Berne Convention for the Protection of Literary and Artistic Works. Copyright law in Europe, Japan, and the United States provides enforceable ownership interests for the authors of computer programs. The scope of that protection varies substantially among the different countries of the world, however.

Copyrights for computer programs in Europe are governed primarily by the European Community Directive 91/250/EEC of May 14, 1991, on the legal protection of computer programs. Under the terms of that directive, computer programs are protected for a term equal to the lifetime of the author plus 50 years. For software controlled by a corporation or party other than the author of the software, the copyright term is 50 years from the date on which the software was first made available to the public. European copyright law protects both the source code and the object code for computer programs. Source code consists of program statements written in a form that can be read by humans, using a computer programming language such as C. Object code is computer program code written using ones and zeros in a form readable by the computer. People write source code, which is then compiled and assembled into object code, which functions as the executable code.

Copyright law in Europe provides the developer of software the exclusive right to reproduce, translate, adapt, alter, and distribute the software he or she creates. The law permits a licensee of software to create a back-up copy of the work without obtaining additional authorization from the copyright owner. In Europe, decompilation of software is permissible to the extent that it is necessary in order to facilitate interoperability, provided that the information derived from the decompilation was not previously made available to the public. This provision was included largely because of the concern that unrestricted copyright ownership could be used to impede interoperability among diverse software products and systems.

In Japan, the Japanese Copyright Act, as amended in 1988, provides copyright protection for computer programs. Japanese copyright law protects software for a term equal to the lifetime of the author plus 50 years or for a period of 50 years from the date of the first public release of the software, whichever term is longer. Japanese copyright law protection for computer programs is interpreted to apply both to the source code and to the object code (Taito Corp., Tokyo District Court, 1982).

An important issue in interpreting Japanese copyright law in the context of computer software is the issue of "creativity" versus "originality." Japanese copyright law focuses on protection of creative works and provides less protection for material deemed to be the result of less creative effort. This approach differs somewhat from that applied in the United States, for example. In the United States, the focus of copyright protection is the originality of the work. Work found to be original in nature (i.e., the result of independent creative effort by the author) is eligible for copyright protection. In Japan, originality of a work is not sufficient. Japanese law does not equate originality with creativity; thus, in a copyright dispute, a Japanese court makes a determination of the creativity of the material in question.

Courts in Japan have examined the issue of creativity with regard to computer programs. For instance, a Japanese court concluded that a computer program that operated a printer was an algorithm that did not display sufficient creativity to merit copyright protection (*System Science v. Toyo Sokki*, Tokyo District Court, 1987). In another case, the court determined that certain batch files integrated into a program were data files, not creative works, and were thus not protected by copyright law (*IGM v. Mets*, Tokyo District Court, 1994). To be protected by Japanese copyright law, computer programs must be creative works; thus, developers seeking copyright protection in Japan should develop a record of evidence that demonstrates that their programs are both original and innovative, the result of creativity by the developer.

Japanese copyright law also specifically excludes some important software elements from protection (Japanese Copyright Act, Article 10[3]). For example, programming languages and rules are excluded from copyright protection in Japan. Algorithms incorporated into programs are also excluded. As a result of these limitations, user interface specifications are not eligible for copyright protection in Japan. In addition, Japanese copyright law restricts decompilation of computer code. These exclusions are motivated, in part, by a desire to ensure that interoperability among different software systems is preserved.

In the United States, the Digital Millennium Copyright Act (DMCA) modified prior United States copyright law to expand the copyright protection afforded to digital content, including computer programs (17 United States Code, Section 1202). Copyright law in the U.S. provides protection for a term of the lifetime of the author of the work plus 70 years. If the copyright of a work is controlled by a corporation or party other than the author, the term is the shorter of 95 years from the date of first publication or 120 years from the date of creation (17 United States Code, Section 102).

One issue that has a significant impact on U.S. copyright law applied to computer programs is the scope of that protection. Under U.S. law, copyright protection is provided to the programming code that creates software; however, that protection does not extend to the "look and feel" of the software. This approach has been applied by the majority of U.S. courts in cases such as *Lotus Development Corporation v. Borland International, Inc.* This means that, although U.S. copyright law permits a software developer to protect the actual code used to create a program, the law is far less effective at enabling the developer to protect the on-screen appearance generated by the program.

Another important aspect of U.S. copyright law is the anticircumvention provision established in the law by the DMCA. This provision in the DMCA provides for both civil and criminal law penalties for developers, distributors, and users of technology (including computer programs) that can be used to "circumvent" systems designed to protect copyrighted material (e.g., software, music recordings, video content) from unauthorized use. The penalties of the anticircumvention provision apply regardless of whether the circumvention technology has actually been used to make unauthorized use of content and regardless of whether that unauthorized use caused any actual harm to the owner of the protected material. In addition, the provision can be applied to technology that merely has the capability of facilitating unauthorized use of copyrighted material, regardless of whether it also has other lawful applications.

There are several examples of application of the anticircumvention provision of the DMCA to different computer programs. For example, software that enables peer-to-peer file sharing has been successfully challenged under the DMCA (*RIAA v. Napster*). Similar successful challenges were raised against developers and distributors of the DeCSS program that supports decryption of DVD content (*Universal Studios v. Reimerdes*, and *Universal Studios v. Corley*). Challenges based in part on the anticircumvention provision are also pending against various other peer-to-peer software systems, including Aimster (*Zomba Recording Corp. v. Deep*), Scour (*Twentieth Century Fox Film Corp. v. Scour, Inc.*), and Morpheus (*MGM Studios, Inc. v. Grokster Ltd.*). In each of these cases, one of the key claims in the complaint is that the computer program in question can be used to circumvent a system designed to protect copyrighted material (e.g., an encryption system); thus, the program is a violation of the DMCA and cannot be manufactured, distributed, or used, even though the program may have other lawful uses. The legal and public policy challenges presented by peer-to-peer systems present a critical set of issues, and those issues are discussed more fully in Chapter 7.

The anticircumvention provision of the DMCA has also been applied to software developers who created their work outside of the United States. For instance, Dimitry Sklyarov, a Russian software developer, was arrested by U.S. authorities while he was in the United States, and was charged with violation of the DMCA (*U.S. v. Sklyarov*). The charges raised by the U.S. authorities were based on computer programs he developed and distributed for use to help access and duplicate electronic book content.

The anticircumvention prohibition included in the DMCA is controversial. Proponents argue that it is a necessary tool to help owners of digital copyrights to enforce their legitimate interests in the networked environment. Opponents of the provision contend that it significantly expands copyright principles in ways that will undermine technical innovation and legitimate user access to copyrighted material. One concern, in particular, receives significant attention from DMCA critics. There is concern that the anticircumvention provision of the DMCA bars, in effect, fair use and legitimate reverse-engineering.

To the extent that the DMCA is interpreted to prohibit use of copyright protected material in ways that can lead to the defeat of copyright protection systems, the DMCA has been used to block noncommercial use of content (e.g., computer programs) that might have qualified as fair use in the past. Traditionally, the principle of fair use made it possible for parties to examine and analyze copyrighted material to conduct research on the material in order to understand how it functions. Properly conducted, that analysis was the basis for legitimate reverse-engineering activities. If the anticircumvention provision is broadly applied, it could make even legitimate reverse-engineering illegal, to the extent that the work under analysis included a copyright protection component or function. Opponents of the DMCA provision contend that this barrier to reverse-engineering could impede the rate and scope of innovation in fields such as computer software and digital media distribution.

It is likely that the anticircumvention provision of the DMCA will be reexamined by Congress in the future. At present, however, this element of the DMCA has significant potential impact on developers, distributors, and users of software who have the capability of overriding or undermining copyright protection measures. This U.S. prohibition is applicable to parties who have any contact with the United States and can be invoked even if the party who is prosecuted did not actually participate directly in any misuse of copyrighted material. Courts are currently attempting to interpret the scope of the provision, and those cases may have a significant impact on future innovation and economic growth in the computer software and digital media industries.

2.2 Patenting computer programs

Patent protection for computer programs is far less widely accepted than is copyright protection. The United States provides the most extensive patent protection for software. From the perspective of the developer, patents provide some advantages over copyright protection. Patents generally provide broader protection than do copyrights, since patents permit their owners to block development of programs that perform substantially similar functions, even if those other programs are not substantially similar in form to their program. This contrasts with copyright law protection, which generally only protects a work from works that are similar in form. Additionally, patent law protects patent owners from developers who may independently create computer programs, while copyright law does not block the work of another developer who develops similar software independently without examining or consulting the original developer.

Patent law also has some disadvantages, relative to copyright law, from the perspective of a software developer seeking protection for his or her work. The process of obtaining patent protection is slower and generally more expensive than that for copyrights. Copyright protection is largely present automatically upon creation of the protected work, while patent protection must be obtained through an application process. Some works that may qualify for copyright protection will not qualify for patent protection. Also, the term of copyright protection is longer than that provided for patent protection. When a patent is awarded, the specific details of the invention are disclosed to the public; thus, parties who apply for patent protection should recognize that if they are successful, the structure and operations of their invention will be publicly accessible. Some inventors may not want to have such public disclosure, and they may accordingly choose to make use of a different form of intellectual asset protection.

Originally, U.S. patent law did not recognize software patents. In the court case *Diamond v. Diehr* (450 U.S. 175 [1981]), however, the U.S. Supreme Court changed that doctrine and accepted the validity of patents for computer programs. Initially, courts in the United States characterized computer programs as "algorithms," and as such, they viewed the programs as laws of nature, which are not patentable. Laws or principles of nature are not patentable, based on the notion that people do not invent such concepts but instead discover them. In the *Diamond* case, the court accepted a new view of computer programs, concluding that they were part of patentable processes and not merely extensions of natural laws; thus, patent protection was granted.

Computer programs in the United States receive patent protection if they meet the standard patent law criteria. Specifically, they must be novel (i.e., unique and distinguishable from all "prior art"). They must be useful (i.e., be capable of performing a valuable function). Finally, they must not be obvious (i.e., they must be innovations that would not be obvious to a person who has reasonable skills in the field in which the invention operates). Computer programs that can satisfy these basic U.S. patent law requirements are eligible for patent protection.

Patent protection is far less extensive in Europe than it is in the United States. The convention on the Grant of European Patents does not directly provide for software patents. Instead, a 1985 modification to that convention permits patents for program-controlled machines and manufacturing processes. Computer programs that are incorporated into those machines and manufacturing processes can be protected under the patents for the overall machine or process. Software standing alone, however, is not covered by the convention.

Several cases in Europe have started to clarify the scope of the software patent protection in Europe; however, the precise extent of that protection remains uncertain. The European Patent Appeal Board, for example, reversed a national court and granted a patent for a digital image processing system (Vicom Systems Application [1987] Official Journal European Patent Office [OJ EPO] 14). However, the board concluded that a computer program to identify certain spelling errors in text was not patentable (IBM, T65/86 [1990] European Patent Office Review [EPOR] 181). The board reversed a German court and approved a patent for an x-ray device that incorporated a data processing system (Koch & Sterzel, T26/86 [1988] OJ EPO 19). In a case involving a patent for a system to correct the visual display of text lettering, the Board found the system to be unpatentable

(Siemens, T158/88 [1991] OJ EPO 566). The board did, however, grant patents for systems that translate printer control features (IBM T110-90 [1994] OJ EPO 557), process computer memory for financial and inventory management operations (Soler, T769/92 [1995)] OJ EPO 525), and adapt television signals to alter their appearance (BBC, T163/85 [1990] OJ EPO 379).

Patents that incorporate computer programs are awarded in Europe when they are part of a machine or process and when they solve an identifiable technical problem. Programs are deemed to solve a technical problem when they perform functions that involve more than mere mathematical calculation. The European Community is not yet prepared to grant patents for stand-alone software. It is also currently unwilling to grant patents for programs that facilitate mathematical calculations but do not perform an identifiable function. This standard remains uncertain, and it continues to evolve as more software patent cases arise in Europe. There appears to be increasing commercial pressure in Europe to grant more software patents; however, strong opposition to such expansion exists as well.

Japan provides somewhat more liberal patent protection for computer programs than does Europe; however, the protection granted there remains less extensive than the patent protection afforded in the United States. Japanese patent law applied to software was originally established in the Japanese Examination Standard for Computer Program Related Inventions (April 1971), and that standard was supplemented by guidelines issued in 1993.

Under Japanese patent law, computer programs are patentable if they are integrated with computer hardware. The programs are also patentable if they establish a cause and effect relationship that is based on a physical law of nature. Thus, as in Europe, if a computer program is integrated into a patentable device or process, the program can be protected by the patent for that device or process. However, unlike European patent law, Japanese law also provides for limited protection for computer programs on a stand-alone basis, but only if the programs implement a function that is based on a physical law of nature. This appears to mean that stand-alone software can receive patent protection in Japan to the extent that it is deemed to apply a natural law or principle; however, this standard has not yet been well defined. Interpretation of the Japanese software patent law remains highly unclear at present.

Some members of the software industry advocate expanded use of patents for computer programs. Other members of that community, however, vigorously oppose award of more software patents. Proponents of computer program patents generally contend that patent protection provides economic incentives necessary to encourage developers to continue to create new software. Opponents of those patents fear that they will reduce innovation in the industry by making it more difficult for developers to modify and build upon code that has already been developed. Widely accepted in the United States, this debate regarding computer program patents is presently most active in Europe and Japan. Opponents of software patents hope to persuade governments in Japan and Europe to limit those patents.

2.3 Business method patents

An intellectual property challenge that is, in part, an offshoot of the rise in popularity of software patents is the question of the role of business method patents. Software applied to support and facilitate business operations and practices (e.g., fund allocations, billing processes, investment decisions) is increasingly protected by patents in the United States. Developers of that software now commonly seek patent protection that extends beyond the programming code and also applies to the business practices that are automated by the software. The key court case accepting the principle of business method patents in the United States was *State Street Bank and Trust Co. v. Signature Financial Group, Inc.* In that case, a federal appeals court reversed a trial court and determined that business methods that make use of mathematical formulas can be patented.

Business method patents are commonly granted in the United States. Other countries, however, do not yet grant business method patents. The debate about the appropriateness of business method patents is an active one in many different parts of the world. Proponents of business method patents contend that they are simply based on the well-established principle of granting patent protection for processes. Viewed from this perspective, business methods are commercial processes that are analogous to the manufacturing and other processes that have long been eligible for patent protection. Opponents of business method patents are concerned that these patents will impede commercial activity, and that the patent examiners who are required to review business methods are not fully qualified to evaluate those processes. There is little dispute as to the appropriateness of granting patents for the software that automates business practices;

however, there is substantial dispute regarding the appropriateness of permitting those patents to extend to the underlying business practices that the software automates.

The business methods patent debate is extremely vigorous with respect to patents for electronic commerce methods. Although there is increasing recognition that computer programs that enable e-commerce functions (e.g., electronic payments, management of online advertising) should be eligible for patent protection, there is growing concern about the grant of patents that cover the e-commerce business methods supported by the software. Online business methods have the same patentable status in the United States as do traditional business methods. This condition has caused disputes in the e-commerce community, with the owners of the e-commerce business method patents contending that the patents are necessary to promote continuing innovation, while other members of the e-commerce community argue that the patents are overly broad and will impede the growth of e-commerce activities.

The leading case on business method patents in the context of electronic commerce is the litigation between Amazon.com and Barnesandnoble.com. The patent at issue in that case is an Amazon.com patent for a "one-click" online payment system. The trial court in that case issued a preliminary injunction in favor of Amazon, requiring Barnesandnoble.com to stop using its similar payment system pending resolution of the full merits of the case. On appeal, the Federal Circuit vacated the injunction and sent the case back to the trial court for resolution of the issue of the validity of the Amazon patent. The appellate court found that Barnesandnoble.com had raised enough issues associated with the validity of the patent to require the trial court to resolve the validity issue. Since the trial court in this case evaluates the validity of the patent, it is likely to provide some clearer guidance regarding standards for e-commerce business method patents.

Another example of a business method patent dispute in the context of e-commerce transaction processing was the litigation between Priceline.com and Expedia.com. That patent dispute was centered on an online reverse-auction process patent obtained by Priceline.com, and the case also involved Expedia.com's owner, Microsoft. A settlement agreement was reached in 2001, before the trial was completed. As part of the settlement, Expedia.com agreed to enter into a license with Priceline.com for use of the reverse-auction system.

The debate over business method patents in both traditional and e-commerce business settings will likely remain active for the near future. In the United States, this debate is taking the form of litigation challenging some of the business method patents that have already been issued and review, by the U.S. Patent and Trademark Office, of the process for granting those patents. In other countries, the debate takes the form of the policy discussion over whether or not those jurisdictions should follow the U.S. model and award business method patents or take a more limited view of those rights. To sample some of the different perspectives on the appropriateness of software patents, read material provided by organizations such as the Software Patent Institute (www.spi.org) and the many different groups in opposition to software patents (e.g., www.freepatents.org).

2.4 Trade secrets protection for computer code

Most developed countries provide legal protection for trade secrets, knowledge, or information that has economic value as a result of the fact that it is secret. The European Community, Japan, and the United States all provide legal protection for trade secrets. Unauthorized use of trade secrets can result in both civil and criminal law liability. Trade secrets are commonly defined very broadly. They may include plans, information, or virtually anything that provides its owner with a commercial advantage over competitors as a result of the fact that the owner has access to the secret but the competitors do not. Accordingly, trade secrets can include computer programs or the expertise necessary to develop a program, and numerous court cases support the application of trade secrets protection to computer code (*Rivendell Forest Products Ltd. v. Georgia-Pacific Corp.* and *Barr-Mullin, Inc. v. Browning*).

To qualify for protection, confidential information must be treated as important secrets by their owners. This means that the owner of the secret must apply a reasonable level of security to protect the information. Those security measure commonly include clear identification of trade secrets, effective use of contracts to secure the secrets (e.g., confidentiality or nondisclosure agreements), and use of physical security measures to manage access to, and use of, confidential material. Failure to provide reasonable security can result in the loss of trade secrets protection. If an owner of a trade secret fails to provide reasonable security to protect the secret, that owner may lose both the secret itself and the legal right to seek compensation for the loss. There is no set term for a trade secret; it can be effective for as long as the protected material continues to have economic value and to be treated as confidential. Perhaps the most famous example of an effective trade secret is the formula used by the Coca-Cola Company for its product, a trade secret that has been in place for many years.

Parties who misuse the trade secrets of others can face private legal actions brought by the owner of the secrets. The most common claim in this setting is a lawsuit for misappropriation of trade secrets. These trade secrets claims are commonly expressed in lawsuits claiming unfair competition or unlawful trade practices. The principle behind this approach is that theft or other misuse of trade secrets represents a commercial practice that is both inappropriate and harmful to commerce.

Courts in the United States determined that a computer program can qualify for trade secrets protection even if it contains some components that are common knowledge in the industry (*Computer Care v. Service Systems Enterprises, Inc.* and *Harbor Software, Inc. v. Applied Systems, Inc.*). Courts in the Unites States have also been willing to find misappropriation of a trade secret even if the infringing software used the secret material in a manner not identical to the owner's application (*Vermont Microsystems, Inc. v. Autodesk, Inc.*). Courts in the United States have also determined that trade secrets law does not bar reverse-engineering in the context of computer software (*Sony Computer Entertainment v. Connectix Corp.*).

There are also specific statutes in some countries that can apply both civil and criminal law penalties for theft of trade secrets. For example, the United States enacted the Economic Espionage Act (18 U.S.C. Sections 1831–1839) and the No Electronic Theft Act. These federal statutes provide for both private and criminal claims against certain forms of theft of secrets. The Economic Espionage Act focuses on commercial secrets that are misused by foreign parties or their agents. The No Electronic Theft Act focuses on the unauthorized use of computers to steal confidential digital material and provides for penalties even if the thief did not realize any economic gain from the theft.

Many other countries have also enacted criminal statutes barring theft of trade secrets. Denmark's Penal Code, for example, provides for prison terms of up to two years for theft of trade secrets. Germany's Penal Code imposed terms of up to three years in prison for misuse of trade secrets. Through its Criminal Information Law, Portugal imposes terms of one to five years in prison for use of computers in an effort to steal trade secrets.

The European Community is currently implementing the Cybercrime Convention, and that treaty includes private and criminal penalties for the theft of intellectual property. The convention makes some forms of misuse of intellectual property criminal conduct. When in full force, the convention will thus make parties who misuse computer programs of others subject to criminal sanctions in addition to any private legal claims under intellectual property law that they may face.

Unlike copyrights and patents, trade secrets do not have a set term. They do not expire. Legal protection for trade secrets remains in effect for as long as the material has commercial value and is protected by the owner. In addition, trade secrets do not require any form of government authorization. Trade secrets protection can be claimed and enforced without any prior government approval.

One of the difficulties associated with the use of trade secrets, however, is that they do not provide complete protection. They do not, for instance, protect an owner from independent discovery of the secret by a competitor. If confidential material is discovered independently (i.e., without access to the proprietary material), the party who discovers it has full right to use the material. Similarly, if the owner of the trade secret makes it publicly accessible, the protection of trade secrets law is lost. For example, a court in the United States determined that publication of a trade secret on the Internet can destroy trade secrets status (*Religious Technology Center v. Lerma*). Also recognize that if material claimed by one party as a trade secret becomes publicly accessible, through no misconduct by any party, that material can be used by any party for any purpose.

Trade secrets also bear the disadvantage that sometimes, in order to engage in litigation to enforce those rights, secrets may be exposed. Some jurisdictions, including the United States, provide methods through which trade secrets that must be exposed to a court during a lawsuit can continue to be protected from public disclosure. Other jurisdictions (e.g., Japan, Germany), however, do not provide such protections. In those jurisdictions, a party who initiates a lawsuit to enforce trade secrets rights may be required to reveal those secrets in court without the benefit of protection from additional public disclosure. In that setting, the practical value of the trade secrets right is substantially reduced. If commercial secrets may be publicly exposed in the course of enforcing secrecy rights, parties will be reluctant to rely on those rights.

Exercise of trade secrets rights in the context of computer programs can sometimes be limited by other legal principles and rights. For example, a California state trial court determined that efforts to block the online distribution of the DeCSS DVD decryption software, based in part on a theory that the software enabled the theft of trade secrets (the trade secrets being the DVD content), were a violation of U.S. constitutional rights of free expression. On appeal, however, the trial court's action was reversed (*DVD Copy Control Assoc. v. Bunner*). The case does, however, illustrate the principle that trade secrets claims must be applied in conjunction with the full range of other legal rights available to users of the secrets.

2.5 Moral rights

Several members of the European Community, as well as Japan, also recognize legal rights known as "moral rights." Moral rights are essentially the rights that an author of a work has in protecting the creative integrity of the work. They tend to be based on the view of the creator of material as an artist. Where recognized, moral rights are viewed as a supplement to traditional intellectual property rights. They tend to provide the creator of artistic content the right to have an influence as to the form and presentation of the work.

Moral rights focus on creative control instead of economic benefit. They have an artistic focus instead of a commercial one. Accordingly, creators who pursue moral rights claims as to their work commonly seek a court order regarding the presentation of the work or the integrity of the work, not payment for use of the work.

Moral rights are not commonly recognized in the United States, and they have not been widely applied to computer programs. They remain, however, a potential source of legal rights for all content creators, including software developers. They may become more of a factor in software protection strategies, as computer software becomes more commonly associated with digital media presentations. Accordingly, before using or modifying computer code created by another party, one should make sure that moral rights have been accounted for along with the other forms of basic intellectual property rights.

2.6 Licensing

The traditional legal mechanism used by software developers to manage the distribution of their products is the software "license." A license is a contract that grants some portion of the owner's rights to another party. Licenses are used to convey rights of access and use for software regardless of whether the property right at issue is a copyright, patent, or trade secret. When moral rights are involved, some jurisdictions take the position that moral rights can never be extinguished or transferred and that they remain enforceable by the creator of the work forever. In such jurisdictions, a contract arrangement such as a license may not be sufficient to transfer or eliminate the creator's right to protect the integrity of the work.

Licenses for computer software were originally customized contracts negotiated by the developer and the licensee. The rise of mass market software has, however, led to the current environment in which a substantial portion of the computer software now commercially distributed around the world is sold subject to the terms of standardized licenses, which are not the result of true commercial bargaining between the interested parties. Mass market software is widely distributed subject to terms presented in "shrinkwrap" licenses.

There has long been debate over the validity of these standardized, shrinkwrap, licenses. Proponents of those agreements argue that mass market sale of computer software would be impossible without some type of standardized and streamlined license agreement. Opponents of the licenses contend that they are unfair to consumers and that the mass market software products should be sold subject to terms that are known to consumers prior to the purchase and are regulated by consumer protection laws.

Courts were initially reluctant to enforce shrinkwrap licenses (*Step-Saver Data Systems, Inc. v. Wyse Technology* and *Arizona Retail Sys., Inc. v. The Software Link, Inc.*). This reluctance was based on the concern that shrinkwrap licenses do not reflect effective bargaining between the seller of the product and the buyer. Sealed in the product and inaccessible to the buyer until after the product has been purchased, courts were concerned that buyers could neither influence the terms of the agreement nor even be fully aware of the terms of the agreement prior to purchase. Courts commonly consider this type of agreement a contract of adhesion, a contract that is presented without negotiation by a seller who has much greater bargaining power than the buyer. Courts generally do not like to enforce contracts of adhesion, since they fear that those agreements take advantage of consumers.

As mass market software has become more widespread, however, courts have been willing to enforce shrinkwrap licenses, subject to specific requirements (*ProCD, Inc. v. Zeidenberg*). Courts now commonly enforce shrinkwrap licenses provided that the terms of the licenses are reasonable and are consistent with terms that have now become generally accepted practices in the consumer software marketplace. Shrinkwrap licenses that contain terms that are unreasonable or are inconsistent with industry norms may not be enforced. Enforcement of shrinkwrap licenses containing commercially reasonable terms is based, in part, on a recognition that computer software is not a consumer product subject to essentially standard licensing terms that both sellers and buyers have come to understand and to accept. That enforcement is also based, in part, on a desire to encourage increased expansion of the commercial marketplace for computer software. Courts remain, however, willing to use the contract of adhesion theory to refrain from enforcing shrinkwrap licenses in those cases where the terms of the license are unreasonable, when compared with standard industry licensing terms.

Some jurisdictions are moving to enact statutes that will clarify the acceptable scope of shrinkwrap licenses. For instance, a standardized statute, the Uniform Computer Information Transactions Act (UCITA), has been enacted in two states (Virginia and Maryland) in the United States and is being considered in several others. Among other actions, UCITA accepts the validity of reasonable shrinkwrap licenses. The E-Commerce Directive under consideration by the European Community (EC Directive 2000/31/EC of the European Parliament and the Council of June 8, 2000) calls upon member countries to adopt laws that will facilitate the expansion of electronic commerce in the European Community. One aspect of that effort could involve more widespread acceptance of shrinkwrap and other mass market licenses for computer software.

In some instances, there can be a conflict between negotiated software license terms and basic intellectual property rights. In those cases, legal action commonly involves both contract law claims and intellectual property law arguments. When the terms of a software license are in dispute, the legal remedies available are commonly those of both contract law and intellectual property law. One highly visible example of this dual nature of software licenses was the licensing dispute between Microsoft Corporation and Sun Microsystems as to the Java programming language. That dispute went to court as both a copyright and a breach of contract action (*Sun Microsystems, Inc. v. Microsoft Corp.*). Sun objected to the Java-compatible products that Microsoft had developed, claiming that those products involved misuse of copyrighted material controlled by Sun. Microsoft contended that its products complied with the modification rights granted in the license it had obtained from Sun. This dispute was ultimately settled; however, it illustrates the extent to which the licensing process commonly integrates contract law rights, responsibilities, and remedies into virtually all intellectual property arrangements.

2.7 Reverse-engineering

Reverse-engineering is the process of analyzing intellectual property that is lawfully acquired with the intent of determining how the property operates. After such analysis, the acquiring party can, under certain circumstances, use the knowledge it derived from the analysis to create its own version of the property. Reverse-engineering is a legitimate and well-established technique in the software marketplace; however, there are limitations on its use.

Reverse-engineering rights may be limited by contracts. License agreements for software commonly include restrictions on reverse-engineering. Those contract limitations are enforceable by the licensor. One possible set of limitations on the ability of parties to waive reverse-engineering rights by contract arises when there is unequal bargaining power between the parties. If the licensor has substantially greater bargaining power than the licensee, and if the prohibition on reverse-engineering has an adverse effect on commercial competition, courts may be more reluctant to enforce the prohibition.

Reverse-engineering for the purpose of ensuring interoperability between products created by different developers is a widely accepted practice. The European Community Directive on the legal protection of software, for example, expressly grants software licensees the right to decompile licensed software for the purpose of maintaining interoperability between their software and the licensed product. Authorities will likely continue to be sympathetic to reverse-engineering applied to foster greater system compatibility and enhanced commercial competition.

2.8 Comparing traditional protection strategies

Copyright and trade secrets laws have been the most widely applied vehicles of intellectual property law protection for computer programs. These forms of legal protection have been most widely used, to date, simply because they were the easiest for developers to invoke. Copyright and trade secrets rights can be established through the actions of the owner of the material; they do not require initial government approval, and this makes them an attractive mechanism for creators of software.

To the extent that patents are now commonly granted for software in the United States, that form of protection has rapidly grown in popularity. As noted previously, patents are attractive to developers looking for protection, since they generally grant a wider scope of protection than do copyrights or trade secrets. The disadvantages with patent protection include the expense and length of time required to obtain protection. Indeed, the amount of time necessary to obtain a patent may make it an unreasonable vehicle to the extent that the software in question has a relatively short useful lifetime. In addition, software patents have incited controversy based on the fear that they will make software less accessible for users.

For the foreseeable future, expect software developers to use a combination of these traditional legal rights to manage their work. Copyrights and trade secrets will continue to be the vehicle of protection of choice for most forms of software. Patent protection will likely become increasingly popular for computer programs that have longer commercial life spans and required more significant investment in their development. Patents are also likely to become more popular for developers seeking protection if they continue to be awarded for both underlying software and associated business methods (both traditional business methods and electronic business methods).

The choice of legal theories applied to control access to software will have an important impact on all users of the software. In general, copyright and trade secrets laws provide more clearly established principles to preserve reasonable access to protected content. For example, copyright law makes use of the well-established principle of fair use to help ensure access to copyrighted material. Trade secrets law recognizes the practice of reverse-engineering and permits that practice, subject to certain limitations.

As patent law becomes increasingly involved in the protection of computer programs, patent law practices may need to be examined to ensure that necessary access to patented material will become more common in the context of software. For example, principles such as compulsory licenses have long been a part of patent management. The principle of compulsory licenses places a legal obligation on patent owners of certain forms of technology (e.g., some types of medical equipment or drugs) to make their technology available to all who want it, subject to fair and reasonable license terms. As the number of software patents increases, we will likely see more widespread application of compulsory licenses as a means to ensure access to patented software, to the extent that such access is necessary for the broader public interest in continuing invention and innovation.

Expect all of the traditional forms of intellectual property law to work to accommodate both the rights and needs of software developers and those of software users. Governments increasingly recognize the need to provide intellectual property protection to ensure that necessary economic incentives for developers exist, as well as the need to provide technical interoperability and dynamic commercial competition by ensuring user access. In that setting, the traditional methods of protecting ownership of software assets will continue to be widely accepted. In addition, however, principles of fair use, open licensing, and fair competition will likely be more actively applied to balance the ownership interests of developers with the need for access by users.

Selected bibliography

A&M Records, et al. v. Napster, 2001 U.S. App. LEXIS 1941.

Amazon.com v. Barnesandnoble.com, Case No. 00-1109 (Fed. Cir. Feb. 14, 2001).

Apple Computer, Inc. v. Microsoft Corp., 35 F. 3d 1435 (9th Cir. 1994).

Arizona Retail Sys., Inc. v. The Software Link, Inc., 831 F. Supp. 759 (D. Ariz. 1993).

Autoskill, Inc. v. National Educ. Sup. Sys., Inc., 994 F. 2d 1476 (10th Cir. 1993).

Barr-Mullin, Inc. v. Browning, 108 N.C. App. 590, 424 S.E. 2d 226 (1993).

Berne Convention for the Protection of Literary Works of Sept. 9, 1886, as amended in 1896, 1908, 1914, 1928, 1948, 1967, 1971, and 1979.

Computer Assoc. Int'l., Inc. v. Altai, Inc. 982 F. 2d 693 (2d Cir. 1992).

Computer Care v. Service Systems Enterprises, Inc., 982 F. 2d 1063 (7th Cir. 1992).

Convention on the Grant of European Patents, at <www.european-patent-office.org/epc/e/ma1.html>.

Criminal information Law of Portugal, Chapt. 1.

Data East USA v. Epyx, Inc., 862 F. 2d 204 (9ᵗʰ Cir. 1988).

Diamond v. Diehr, 450 U.S. 175 (1981).

DVD Copy Control Assoc. v. Bunner, 113 Cal. Rptr. 2d 338 (Cal. Ct. App. 2001).

European Council Directive 91/250/EC of May 14, 1991, on the legal protection of computer programs, available at <http://europa.eu.int/eur-lex/en/lif/dat/1991/en_391L0250.html>.

European Council Directive 2000/31/EC of June 8, 2000, on certain aspects of the information society services, in particular electronic commerce in the internal market, available at <http://europa.eu.int/ISPO/ecommerce/legal/documents/2000_31ec/2000_31ec_en.pdf>.

Feist Publications, Inc. v. Rural Telephone Service Company, 499 U.S. 340 (1991).

German Penal Code, Sect. 202.

Harbor Software, Inc. v. Applied Systems, Inc., 1996 U.S. Dist. LEXIS 13224 (S.D. N.Y. Sept. 9, 1996).

IGM v. Mets, Tokyo Dist. Ct. (1994).

In re BBC, T163/85 (1990) OJ EPO 379.

In re IBM, T65/86 (1990) EPOR 181.

In re IBM, T110/90 (1994) OJ EPO 557.

In re IBM, T854/90 (1993) OJ EPO 669.

In re Koch and Sterzel, T26/86 (1988) OJ EPO 19.

In re Petterson, T1002/92 (1995) OJ EPO 605.

In re Siemens, T158/88 (1991) OJ EPO 566.

In re Soler, T769/92 (1995) OJ EPO 525.

In re Vicom Systems (1987) OJ EPO 14.

Lotus Dev. Corp. v. Borland Int'l., Inc., 49 F. 3d 807 (1ˢᵗ Cir. 1995), affirmed by the U.S. Sup. Ct., 116 S. Ct. 804 (1996).

Metallica, et al. v. Napster, Case No. 00-13914 (S.D. Cal. filed Apr. 13, 2000).

MGM Studios, Inc., et al. v. Grokster Ltd., et al. (W.D. Cal. filed Oct. 2, 2001), Complaint available at <www.riaa.com/pdf/complaint.pdf>.

Penal Code of Denmark, Sect. 263.

Priceline.com v. Expedia.com (D. Conn. Filed Oct. 2001), settlement agreement executed, January 2001.

ProCD, Inc. v. Zeidenberg, 86 F. 3d 1447 (7ᵗʰ Cir. 1996).

Religious Technology Center v. Lerma, 1995 U.S. Dist. LEXIS 17833 (E.D. Va. Nov. 28, 1995).

RIAA v. Napster, et al. v. Napster, 2000 U.S. Dist. LEXIS 5761.

RIAA, et al. v. Napster, Case Nos. 00-16401 and 00-16403 (9th Cir. 2001).

Rivendell Forest Products Ltd. v. Georgia-Pacific Corp., 28 F. 3d 1042 (10th Cir. 1994).

Sony Computer Entertainment v. Connectix Corp., 203 F. 3d 597 (9th Cir. 2000).

State Street Bank & Trust Co. v. Signature Financial Group, Inc., Case No. 96-1327 (Fed. Cir. July 23, 1998).

Step-Saver Data Systems, Inc. v. Wyse Technology, 939 F. 2d 91 (3d Cir. 1991).

Sun Microsystems, Inc. v. Microsoft Corp., Case No. 97-CV-20884 (N.D. Cal. 1998).

Sun Microsystems, Inc. v. Microsoft Corp., Case No. 99-15046 (9th Cir. 1999).

System Science v. Toyo Sokki, Tokyo Dist. Ct. (1987).

Taito Corp., Tokyo Dist. Ct. (1982).

Twentieth Century Fox Corp., et al. v. Scour, Inc. (S.D. N.Y. filed Jul. 20, 2000), complaint available at <www.mpaa.org/Press>.

Universal Studios, et al. v. Corley, Case No. 00-9185 (2d Cir. 2001).

Universal Studios, et al. v. Reimerdes, Case No. 00 Civ. 0277 LAK (S.D. N.Y. Aug. 30, 2000).

U.S. Copyright Act, 17 U.S.C. Sect. 102.

U.S. Economic Espionage Act, 18 U.S.C. Sects. 1831-39.

U.S. Uniform Computer Information Transactions Act (UCITA), adopted by the National Conference on Commissioners on State Laws on July 29, 1999, information available at <www.ucitaonline.com>.

U.S. Uniform Trade Secrets Act, 12 U.L.A. 433.

U.S. v. Sklyarov, Case No. 5-01-257 (N.D. Cal. Jul. 10, 2001), complaint available at <http://cryptome.org/usa-v-sklyarov.htm>.

Vermont Microsystems, Inc. v. Autodesk, Inc., 1994 U.S. Dist. LEXIS 18737 (D. Vt. Dec. 23, 1994).

Zomba Recording Corp., et al. v. John Deep, et al., Case No. 01 CV 4452 (S.D. N.Y. filed May 24, 2001), Complaint available at <www.riaa.com/pdf/aimster_complaint.pdf>.

3

Open Source: An alternative model

In the previous chapter, we discussed the traditional legal methods applied to protection and management of computer programs. This chapter will focus on a software development model that has been in existence since the beginning of the computer age, but which has only relatively recently achieved a state of widespread commercial acknowledgment. The "Open Source" development model for computer programs now has significant impact on the creation, distribution, and use of computer software in the commercial marketplace. Source code, as we noted in the previous chapter, is the computer programming statements that are readable by humans and written in various programming languages. The Open Source distribution system provides a mechanism for sharing access to source code among many different software developers. Open Source distribution has long existed in the software world; however, it was not generally applied as one of the traditional software management processes. That situation has changed, and Open Source software now plays a substantial role in the computer industry. It is likely that the role of Open Source material will continue to increase in the future. This chapter provides an overview of the Open Source model and presents some ideas as to strategies for effective software management in an Open Source environment.

Access to source code plays a significant role in commercial software development. If source code is made more accessible, diverse applications compatible with the original source code tend to develop more quickly. Examining operating systems, for example, we see that the Linux operating system has been able to facilitate rapid development of diverse applications to run with that system largely because Linux has been distributed on an Open Source basis. Contrast that approach with the distribution strategy applied by Microsoft for its Windows operating system, which relies on a traditional proprietary system that blocks access to the source code. Open Source access to an operating system's source code facilitates development of applications to run on the operating system by making it easier to develop and use Application Programming Interfaces (APIs). APIs are the software that application programs use to communicate and work with operating system programs. The Open Source strategy involves a commercial trade-off. It is a system that foregoes, at least in the short term, licensing revenues in exchange for the opportunity to make the product a widely accepted industry standard, or platform, and thus set the stage for greater revenues from multiple sources in the future. In this chapter we discuss some of the key issues associated with that trade-off.

3.1 Open Source principles

At its heart, the Open Source development model is based on the premise that widespread access to the underlying source code for computer programs carries greater overall public benefits than does proprietary control over that material. Proponents of the Open Source approach to software development emphasize a few key advantages of the process. They note that the system facilitates rapid development of improvements for the Open Source software. Open access to the source code by many different developers enables those developers to correct deficiencies in the program quickly. That access also enables developers to create enhancements and new applications for the original code more quickly than they could without such access.

The more rapid development of enhancements and applications encouraged by the Open Source process increases the pace of market penetration for the original program as well. Advocates of Open Source development thus contend that the process helps a software developer to refine and enhance its product more quickly, at lower cost, than would be the case if applying a strict proprietary development model. Viewed from this perspective, Open Source development provides a more efficient and economical means of supporting software innovation.

In its purest form, the Open Source development process provides an efficient mechanism enabling licensees of software to customize the software for optimal use when applied to the specific functions the licensee has in mind. The simplest form of Open Source development provides a process through which licensees of software can modify the code to ensure that the software functions effectively for their specific purposes. Open Source was initially intended to serve as a system allowing the party best positioned to understand the application needs of the software (the licensee) to have the authority to alter the source code as necessary to meet those needs effectively.

3.2 Open Source and traditional intellectual property principles

The Open Source development process simultaneously challenges traditional property rights associated with intellectual property and relies upon those rights. The Open Source model challenges traditional intellectual property management strategies by providing greater access to that property than is generally afforded under conventional licensing strategies. Yet the model also relies on standard intellectual property law principles by applying those principles to enforce the terms of the Open Source license. Accordingly, Open Source development systems simultaneously undermine and enhance traditional theories of intellectual property management.

Software distributed under Open Source terms continues to be intellectual property subject to legally enforceable licenses. Some parties confuse the liberal licensing terms of Open Source with relinquishment of intellectual property ownership rights. This is not the case. A software developer who distributes software using Open Source terms is licensing that software, not surrendering ownership. Open Source material is not part of the public domain; instead, it continues to be the property of the developer, even though the developer has made the property available subject to liberal licensing terms. An Open Source license is as enforceable as any conventional software license. Open Source licenses, in effect, grant the licensee the right to create "derivative works" based on the original source code in exchange for the licensee's promise to provide a specified level of open access to the modified product. Open source licensors commonly charge no license fee; however, they often require the licensee to engage in some form of conduct in exchange for the zero fee. The Open Source process relies upon a grant of the right to make derivative products combined with retention, by the licensor, of control over the original copyright and retention of some level of control over the derivative work.

There is potential for conflict when different versions of an Open Source product are not fully compatible. In such situations, the derivative work is not compatible with the original product and with some of the other derived works. An example of this type of dispute surfaced in the conflict between Sun Microsystems and Microsoft over competing versions of the Java product. Sun argued that Microsoft had created a version of Java that was not fully compatible with the Sun version of Java. Microsoft claimed that its version of Java was fully consistent with the terms of its license with Sun. The parties ultimately settled this dispute, but the case illustrates the problems that can arise when derivative versions of an Open Source product are not fully compatible.

Copyright law has traditionally been the form of intellectual property rights most widely applied to computer programs; however, the Open Source model also has some interesting implications for software patents. As discussed in Chapter 2, software developers rely increasingly on patent protection to enforce rights of ownership over their work. In order to obtain a patent, an inventor must establish that the invention effectively extends beyond the current scope of the prior art relevant to the invention. Prior art includes previously issued patents and publicly disclosed material. In some instances, the Open Source process for software development can have an important impact on future software patents.

The Open Source development process provides, in effect, a dynamic environment conducive to rapid expansion of prior art in the field of computer software. Code and applications presented through Open Source arrangements are widely accessible; indeed, that widespread and rapid accessibility is an important reason behind choosing the Open Source model. Once presented to users on an Open Source basis, the material has been publicly disclosed and is now part of the prior art, for patent purposes. In this way, expansion of Open Source distribution creates a corresponding expansion in the prior art relevant to software patents. It thus seems that, as Open Source distribution extends its scope, future ability to obtain software patents may decrease, as more relevant material moves into the recognized prior art. Indeed, some parties now view Open Source code distribution as a form of insurance against future patent claims. Advocates of this strategy encourage widespread release of source code in order to flood the prior art for computer software, thus making it more difficult for developers to obtain patent rights in the future.

However, just as Open Source distribution of source code can make it more difficult for parties to obtain patent rights in the future, so too can preexisting patents pose a barrier to the Open Source system. If specific code is protected by a software patent, that patent would prevent integration of the code into an Open Source product without the permission of the patent owner. As the number of software patents increases, more code falls within the scope of the broader protection that patents provide, and in that environment it may be more difficult to expand Open Source offerings. Software that has been patented can also be distributed on an Open Source basis but only with the permission of the patent owner. Some proponents of the Open Source model now obtain patent protection prior to making their product available on an Open Source basis. As noted previously, however, the process of obtaining a patent generally requires more time and money than does copyright protection; thus, patents for Open Source material are at present somewhat less attractive than traditional copyrights.

3.3 Different forms of Open Source

There are different forms of Open Source licenses now applied in the software marketplace. All forms provide licensees with the ability to access and modify the source code of the original product. The forms differ, however, with regard to the obligations placed on the licensee as to distribution and use of the modified source code. Generally, an Open Source licensee is permitted to modify the source code at no charge and to retain the modifications as proprietary provided that the licensee uses the modified software only for internal purposes, making no distribution of the modified version. The Open Source licensors tend to vary with respect to the obligations they impose on a licensee planning to distribute the modified software.

One form of Open Source license is the "General Purpose License." This license permits the licensee to read and modify the source code of the original product. In exchange for this grant of rights, the licensee must agree to distribute all modifications the licensee makes in the source code on an Open Source basis, if the licensee distributes the modified software.

Under the General Purpose License, the licensee can treat the modified source code as proprietary material only if the licensee does not distribute the software. If the licensee modifies the original software exclusively for internal use, the modification need not be distributed on an Open Source basis. If, however, the licensee chooses to distribute the modified software, under the terms of the General Purpose License the licensee must make the modified source code available on an Open Source basis.

Another basic Open Source licensing model is the "Berkeley Software Distribution" License (BSD). Under the BSD model, the licensee agrees that the copyright for the original software remains fully within the control of the Open Source licensor; however, the licensee is not required to release the source code for modifications developed by the licensee. The licensee is permitted to build upon the original code and keep the modified code proprietary, either for the developer's private use or for future commercial distribution. With this structure, the BSD License is generally considered to be more attractive to commercial developers.

Different variations on the General Public License and the BSD License have also been implemented. For example, Netscape applied the Netscape Public License to its Navigator Internet browser. Under the terms of that form of Open Source licensing, developers were free to create proprietary versions of the base code; however, Netscape retained the right to have access to all source code enhancements made by licensees and to incorporate all of those enhancements into products that Netscape would treat as proprietary. Some parties were critical of the Open Source model initially applied by Netscape, and Netscape elected to modify its Open Source licensing model when it developed the Mozilla Public License for its Mozilla program. Mozilla originated in the Netscape Navigator and Communicator products, and it now consists of a set of Internet client software products developed and distributed on an Open Source basis. Under the terms of the Mozilla Public License, Netscape no longer retained the right to include source code enhancements into Netscape proprietary products.

Different Open Source products are made available subject to different Open Source licensing models. Essentially, all of the different Open Source licensing models are variations of the General Public License or the BSD License. For example, software products such as UNIX, Apache, Sendmail, Solaris, and Mozilla are generally distributed using versions of the BSD License. Linux is distributed using a license that more closely resembles the General Public License. Products such as Perl make use of the "Artistic License," a version of Open Source licensing that generally requires source code modifications to be made available on an Open Source basis but that also permits some developers to retain certain modifications as proprietary and sometimes permits the developers to introduce their modifications into the public domain (i.e., relinquish copyright ownership claims as to those modifications).

3.4 Open Source versus shareware

The Open Source development and distribution model differs significantly from the well-established system of "shareware." Although some observers blur the distinctions between the two approaches to software distribution, they differ substantially. Software distributed as shareware does not generally carry with it access to the underlying source code. Users of shareware products are not, as a matter of course, authorized to view or modify the source code of the shareware. In contrast, Open Source products permit some level of access to the source code. The grant of rights to the licensee for Open Source software is thus significantly greater than the rights of use provided to a licensee of a shareware product. In effect, shareware licensees are granted the right to use the software in question at no charge; however, they are not authorized to modify the source code of the software that has been shared.

3.5 Open Source and Open Systems

At times, there is confusion as to distinctions between the concepts of Open Source and Open Systems. Both principles speak to the issue of promoting broader access to digital material; however, they do differ from each other. The concept of Open Systems is essentially premised on the notion that economic and technological efficiency is best served if computer systems and content provided by different organizations are compatible with each other. An Open Systems philosophy promotes ready interconnection and interoperability among hardware and software provided by different parties. The collaborative aspects of the Open Source process fit well with an Open Systems philosophy; however, a formal Open Source approach is not necessary to facilitate Open Systems. Opens Systems can be supported through enforcement of rules requiring operational consistency and compatibility, even if there is no actual grant of Open Source access to the underlying source code of the products in question.

3.6 Economic models for Open Source material

Although many of the benefits identified by proponents of Open Source do, in fact, exist, there are also some disadvantages associated with that development process. For example, the Open Source process generally makes it more difficult for the original developer of the code in question to derive revenue directly from distribution of the code. Traditional revenue streams from software licensing are dramatically reduced when the distribution system applied is a version of Open Source. At present, there is substantial uncertainty as to the most effective economic model for distribution of Open Source material. Traditional intellectual property models are based on revenues derived from licensing. Under the traditional model, copyrights, patents, and other forms of intellectual property laws are used to establish and enforce ownership rights. Revenues are obtained from licenses that are required before parties other than the owner can make use of the material.

Open Source content does not rely on licensing revenues. As noted previously, Open Source content is commonly licensed subject to a zero or nominal license fee. Consideration for the license takes the form of mandatory conduct (e.g., mandatory release of modifications to source code on an Open Source basis) instead of payment of money. The licensor of Open Source products tries to derive commercial benefits other than money from the broader distribution inspired by Open Source distribution.

Without licensing revenues, the traditional revenue source for a software developer is unavailable. Open Source software distributors must, accordingly, look to other sources of revenue to compensate for the reduction in licensing fees. One common revenue source is a charge associated with customization services. Some developers of Open Source products make those products available at no charge but assess fees for the development of customized versions of those products. There is often significant expense associated with refining an Open Source product into a version that can be effectively used by an enterprise; thus, users of Open Source products often turn to outside parties to make modifications to the software appropriate for the enterprise. Parties who want access to the customized versions pay fees for the development of those versions, and those fees can be significant. Enterprises—including RedHat Software and Walnut Creek, for example—provide customized Open Source products and customization services to organizations using Open Source programs.

Another commonly applied model involves assessment of fees for goods or services that are ancillary to the Open Source product. Those ancillary services sometimes include information and operational support for users of Open Source products. Under this approach, the software is made available at low cost, but the licensee pays for other products or services that facilitate the use of the Open Source product by the licensee. Companies such as O'Reilly & Associates provide an example of this type of ancillary service on a fee basis.

Another commonly applied economic model for Open Source products involves the sale of enhanced versions of Open Source products. Under this approach, the developer distributes some of its products on an Open Source basis. Over time, the developer creates enhanced versions of the product, with attractive additional features and capabilities. The developer maintains the base product on an Open Source basis and sells the enhanced products under traditional proprietary licensing models. With this model, enhanced products are eventually distributed on an Open Source basis, as their novelty erodes, and new innovative versions of the product replace them in the proprietary product inventory. Companies such as Sendmail, Inc. and ActiveState Tool Corporation have applied this approach.

Some developers use the popularity of their Open Source products to help them increase revenue from unrelated products or services. For example, at various times Netscape made use of Open Source products such as Mozilla and Communicator to attract increased traffic to its commercial Web sites, thus enhancing revenues from sources such as advertising. This approach uses the Open Source product as a means to attract customers to other revenue-generating products or services. The Open Source product thus serves, at least in part, as a marketing or promotional device.

Ultimately, developers of Open Source products hope to make their products into widely embraced industry standard products. If successful, an Open Source distribution model can create enough market share to establish the product as a platform used by many different customers and encouraging the continuing development of different applications and ancillary programs. In this way, the licensing revenue that was lost during the early stages of Open Source distribution can be more than compensated for by significant revenues derived after the product has become a widely used standard product and market platform.

3.7 The challenge of fragmentation

An important consequence of Open Source distribution of computer code is the rise of diverse versions of the initial program. That diversity presents another potential disadvantage to the developer of the original code. As different developers access and modify the basic source code, multiple "flavors" of that initial code are created. Some have characterized this proliferation of diverse versions of the original code as "fragmentation." If these diverse versions of the code are compatible and function effectively in conjunction with each other, the problems associated with fragmentation can be minimal. In contrast, if some of the different versions do not mesh effectively with each other, fragmentation can have negative technical and commercial consequences.

One example of the sort of widespread product fragmentation resulting from Open Source distribution is provided by the history of the different forms of the UNIX software. Widespread access to the UNIX source code led to the creation and distribution of many different versions of UNIX. Although most of those different versions are compatible, the significant level of diversity for the UNIX product that developed over the years led to some compatibility problems. The fragmentation of the UNIX product led many Open Source developers to pay greater attention to the challenge posed by fragmentation.

Some Open Source developers have tried to address the problem of fragmentation by applying variations of the standard Open Source licensing arrangements. For example, the Apache software is distributed using the Apache license. That license provides access to the Apache source code, but it requires that when modified versions of the Apache code are distributed, they must be identified under names other than Apache. In this way, only the base Apache code is distributed under the name Apache.

Sun Microsystems experimented with another licensing approach to deal with the fragmentation problem. Sun applied its Sun Community Source License (SCSL) to its Java and Solaris products. The SCSL permits developers to review the Sun source code, but it requires that changes to the source code be submitted to Sun for approval before they can be implemented. In this way, only modifications approved by Sun can be incorporated into the base products. The SCSL structure also requires payment of a license fee to Sun when the modified code is distributed commercially, and that provision led some observers to question whether the SCSL model is truly an Open Source system.

The potential for product fragmentation is a seemingly inevitable risk associated with the Open Source approach. Premised on widespread development of modifications by many different developers, the Open Source system invites fragmentation. Indeed, some level of fragmentation is the engine that drives the innovations and enhancements that Open Source development facilitates. Fragmentation is, to a point, both inescapable and desirable. If, however, fragmentation leads to operational incompatibility among the diverse versions of the Open Source product, then it will have adverse consequences for all of the parties involved.

3.8 Open Source applied beyond software

Basic Open Source principles now surface with respect to products in addition to computer software. Increasingly, developers of other creative content (e.g., microprocessors, multimedia content) are prepared to grant other parties the right to create derivative works, based on the original property created by the initial developer. At its essence, this model for shared development is a form of Open Source. Regardless of what the creative content may be, when an initial creator of material is willing to allow other parties to use that material as the basis for modified versions, in exchange for a commitment that the second developer will, in turn, make the modified version available for subsequent modification, there is an open development system substantially similar to the classic Open Source model associated with computer programs.

Challenges and opportunities highlighted by the use of the Open Source model for software distribution provide useful lessons for application of open access principles for knowledge and information assets, as well. Many organizations are now attempting to devise policies and practices that will enable them to manage more effectively the wide range of knowledge assets that those organizations have developed. Some of the practices now applied to management of Open Source software are likely to be applicable for other information and knowledge assets as well.

The Open Source process relies on sharing extremely sensitive and valuable information with a wide range of parties, including those outside of the original developer's organization. Yet, Open Source developers are also looking for ways to retain the necessary level of control over their original product and to derive appropriate revenue from exploitation of that product. The Open Source model has thus forced the parties who apply that model to begin to deal with several issues, including effective management of the assets in a shared access environment and development of viable nontraditional economic models to derive revenue from those shared assets. Those same issues confront organizations that are attempting to implement more sophisticated knowledge management processes for intellectual assets that extend beyond computer software.

As a result, the management processes and economic models now being developed by the Open Source community will likely be applicable for broader knowledge management purposes as well. Practices developed to implement Open Source distribution of software can, in this way, create a valuable template that owners of diverse knowledge and information assets can apply to promote effective utilization of those varied assets. The Open Source marketplace may thus serve as a proving ground for management principles and procedures that will enhance our ability to execute effective knowledge management on a widespread basis. If the Open Source process functions successfully as a proving ground for knowledge asset management strategies, that function may eventually be seen as a contribution to commerce that is far greater than the direct contribution made by the economic value of Open Source software itself.

3.9 A battle for the heart and soul of Open Source

Proponents of the Open Source model do not all agree as to the proper scope and direction of future development for that model. One can essentially divide the Open Source movement into at least two factions. One faction tends to focus on the potential commercial value of widespread implementation of Open Source development processes. The other group tends to emphasize the ideological value of the Open Source process as a means of empowering more individual developers and software users. Both factions are active and passionate. It remains unclear whether either group will ultimately prevail.

Open Source proponents in the first faction tend to emphasize the potential value of Open Source as a means to make software development and distribution more efficient than traditional commercial software development models. They promote the fact that Open Source facilitates faster product development and refinement. They also extol the value of the process as a means of promoting rapid market share expansion. Members of this faction contend that the Open Source model must make room for some level of proprietary control over modified Open Source material if the model is to appeal to commercial developers. They also note that adoption by major commercial software developers is essential if the Open Source process is to provide the product development and distribution efficiency enhancements that it is capable of offering.

Those in the other faction do not, in general, agree that the key strength of the Open Source model is its value as a mechanism to enhance the efficiency of the software development market. Instead, they place greatest value on the potential that Open Source provides to empower software developers and to promote greater interaction among developers. Some have described this movement as the "free software" movement. In this description, the notion of freedom does not refer to the price of access to the software but instead refers to the belief that computer programs should be freely accessible to as many people as possible, for as many different applications as possible. Advocates of free software thus tend to focus on the need to structure and enforce Open Source licenses in ways that minimize restrictions on access to and use of modified code.

Essentially, the more commercially oriented Open Source faction is willing to concede some level of proprietary control over the fruits of Open Source enhancements for the sake of expanding the commercial strength of the overall software marketplace. Members of the free software faction, in contrast, want to move toward elimination of proprietary rights for source code. Essentially, proponents of the free software philosophy believe that computer programs should be widely accessible for any application. To accomplish this, the source code must be available for examination and modification, and the resulting programs must be widely distributed for use by anyone for any function.

It seems fair to summarize the different Open Source philosophies in the following way. Many of the proponents of commercial Open Source focus on the value of that model as an economic and technology development strategy. In contrast, many of the free software advocates emphasize the significance of Open Source as social movement assisting empowerment of the individual. The distinction between these views is significant. The scope of that difference in philosophy makes reconciliation difficult. Emotions in this conflict continue to run high.

3.10 Open Source: Placing a premium on collaboration

Application of the Open Source model carries certain implications for those who would implement that model. Perhaps the most important of those implications is the need to accept a greater degree of collaboration than that required for traditional distribution systems. At its heart, the Open Source development process relies on collaboration. Successful use of the model requires effective collaboration. Effective collaboration often requires business strategies and management skills that differ from those traditionally highlighted in competitive markets. A decision to participate in an Open Source process carries with it the need to commit to effective collaboration with the other participants in that process. Not all individuals or organizations are adequately prepared to thrive in such a collaborative setting.

An inevitable aspect of collaboration is some level of loss of control. When an individual or an organization grants access to assets in a manner such as that applied under Open Source, the developer of that asset cedes some level of control over the asset to its collaborators. Open Source is based on relinquishment of at least some level of control. Effective collaboration requires shared control; thus, the Open Source process is all about diversification of control. To be successful in such collaborative efforts, participants must be comfortable functioning in a setting that is not based on traditional command and control relationships.

3.11 Management strategies in an Open Source environment

There are a few basic management principles and strategies that are important in a market in which the Open Source model could be applied. Those strategies involve recognition of the ways in which Open Source distribution can surface in your industry. They also involve development of methods that will enable your organization to use Open Source effectively, when appropriate, and to respond efficiently to the challenge of Open Source when it is applied by one of your competitors, customers, or business partners.

One important principle is the need to recognize that the Open Source model must be addressed once it surfaces in your market. It is not realistic to assume that you can ignore the implications of the fact that your competitors, customers, or business partners now apply some form of the Open Source model. The need to respond is obvious in situations such as when one of your competitors begins to make some of its products available subject to Open Source licenses. The need to respond is sometimes less clear, however, when, for example, one of your competitors chooses to provide products (e.g., computer equipment) that facilitate or encourage broader use of Open Source materials. It is helpful to recognize that the Open Source model is a force to be reckoned with regardless of how it surfaces in your commercial environment. Some may be tempted to believe that the only decision to be made is whether their organization should make use of the Open Source approach. That view is not entirely accurate. In addition to the decision about whether it will choose to apply the Open Source system to its products, each business should also decide how it will react if another enterprise in its marketplace applies an Open Source strategy.

Another important Open Source strategy concept is the fact that the Open Source model is both disruptive as to intellectual property management and also reliant upon intellectual property management. It is disruptive, since it creates a setting in which the original developer of intellectual assets inevitably relinquishes some measure of control over those assets. It is reliant upon intellectual property management, sine the system depends upon intellectual property law rights in order to function. As noted previously, Open Source does not equate to abdication of property rights. Instead, Open Source is simply an alternative licensing strategy. If an Open Source license is violated, traditional intellectual property law rights present the mechanism through which the violation will be terminated, and compensation will be provided.

Recognize that the Open Source model requires a willingness to explore nontraditional revenue models. Intellectual property and other forms of intellectual assets were traditionally managed through reliance on legal rights to force payment of fees through licensing arrangements. The Open Source distribution model does not fit that traditional revenue generation model well. As we discussed, use of the Open Source model has led developers to apply revenue-generation strategies other than traditional intellectual property licensing. If you make use of the Open Source approach, you will most likely be required to develop and maintain a nontraditional strategy for revenue generation in lieu of classic licensing. Development of an appropriate and effective revenue model should take place before you commit to the Open Source approach.

Appreciate the fact that the Open Source approach to distribution of intellectual assets now affects assets in addition to computer software. For instance, some computer equipment (e.g., microprocessors) and diverse digital content (e.g., new media material) now contains elements distributed subject to versions of the Open Source model. Assume that, in the future, an increasing number of products and services will be directly affected by Open Source distribution. In that setting, prudence suggests that enterprises should anticipate the impact that application of Open Source in their industries would have on their key markets and on their competitive position. Simply because your marketplace has not yet been affected by the Open Source strategy, do not assume that Open Source will never reach it.

Also, remember that some of the economic models and management strategies now being applied to Open Source development can be effective for broader knowledge management initiatives. Organizations are struggling with the desire to share more of their knowledge and information assets with other parties and the need to maintain some level of ownership control over those assets. The management strategies proving to be effective in the Open Source software marketplace may also be effective to promote broader access coupled with retention of necessary control for knowledge assets. Continue to look to the Open Source environment for lessons that can be applied to the challenge of effectively managing knowledge assets in general.

Finally, recognize that there are different Open Source ideologies. The Open Source approach to development and distribution of intellectual assets has a different meaning to different proponents of that approach. As noted previously, some Open Source advocates apply a more commercially oriented form of the system (one that balances broad content access with preservation of certain proprietary rights), while other supporters of the model apply a version of the system that seeks to maximize access while minimizing assertion of proprietary controls. It is likely that both factions will be active for the foreseeable future and that the ideological conflict between them will make management of Open Source products more challenging.

Selected bibliography

Debian Free Software Guidelines, at <www.debian.org>.

Free Software Foundation, at <www.gnu.org>.

Levine, R. et al., *The Cluetrain Manifesto*, Cambridge, MA: Perseus Books, 1999.

Linux International, at <www.li.org>.

Linux Online, at <www.linux.org>.

Mozilla Web Site, <www.mozilla.org>.

Ogg Vorbis, at <www.xiph.org/ogg/vorbis/index.html>.

Open Source Initiative, at <www.opensource.org>.

Raymond, E., *The Cathedral and the Bazaar*, at <www.tuxedo.org/~esr/writings/cathedral-bazaar/cathedral-bazaar/index.html>.

Stallman, Richard, *The GNU Manifesto*, at <www.gnu.org/gnu/manifesto.html>.

Sun Microsystems, Inc. v. Microsoft Corp., Case No. 97-CV-20884 (N.D. Cal. filed 1997). Sun's Complaint is available at <www.java.sun.com/about/java/info/complaint.html>. Microsoft's Answer is available at <www.microsoft.com/corpinfo/java/java2.htm>.

Wayner, P., *Free for All*, New York: HarperCollins, 2000.

Xiphophorus, at <www.xiph.org>.

4

Trademark protection online

Trademark law is particularly important in the online environment. This chapter focuses on some of the most common trademark issues that arise with regard to several key components of online content. It examines the different aspects of trademarks and domain names, hypertext links, and keyword searches. These topics are among the most common ones addressed by trademark law in the online setting. This chapter also summarizes some of the most common additional legal theories frequently raised in conjunction with trademark arguments in the context of domain names, keywords, and links. Those additional legal claims include anticybersquatting regulations and unfair competition claims. The chapter also discusses some of the most common strategies to manage domain names, links, and keywords to minimize liability risks associated with trademark and other legal requirements.

4.1 Trademark overview

Trademarks are commercial identifiers used by businesses to identify their products and services to consumers. They can take virtually any form, and they help the public to distinguish the goods or services of one company from those of its competitors. The most common trademark forms are words, logos, symbols, and music. However, trademarks can take the form of essentially anything that becomes recognized as representation of a specific enterprise or its products or services. One business commonly develops and maintains several different trademarks (e.g., the fast food company, McDonalds, asserts trademark protection over many different marks, including its name, its golden arches logo, and the "Mc" prefix for its products). The key aspect of a trademark is not its form but is instead the fact that it has become synonymous, in the public mind, with a particular company.

Trademark law rights include legal protection for marks associated with both goods and services. Marks associated with products are generally characterized as trademarks. Marks associated with services are commonly referred to as "service marks." Under certain circumstances, distinct forms of packaging that become closely identified in the public mind with a specific company and its products are protected under trademark law as "trade dress." When we talk about trademark law rights, recognize that those rights also apply to properly developed service marks and trade dress.

In most jurisdictions, trademark rights can be established either by registration with the government or by actual commercial use of the mark. Registration is commonly required, however, before the owner of a trademark can enforce its rights against others. In the United States, such registration is processed by the U.S. Patent and Trademark Office, which is a part of the Department of Commerce. To establish trademark rights through use of a mark, the owner of the mark must generally demonstrate that it has consistently used the mark in commerce for an extended period of time. This process is often described as creation of a "common law" trademark right. Not all jurisdictions permit establishment of trademark rights through common law usage, but that approach is permitted in the United States.

Trademark law provides for protection of marks against uses that lead to consumer confusion as to the identity of goods and services ("misappropriation"), and the law also provides protection against uses of marks that undermine the economic value of the marks ("dilution"). The basic legal claim for misuse of a trademark is misappropriation. The claim is also sometimes characterized as infringement. The claim involves an argument that another party has used a mark that is identical, or confusingly similar, to the mark claimed by the party raising the claim. To prevail in a legal claim of misappropriation, the owner of the mark must demonstrate that it has a claim of use for the mark that is superior to that of the defendant. The plaintiff must also persuade the court that the mark used by the defendant is likely to confuse consumers as to the origin of the goods or services that are being sold. If the plaintiff in a trademark action is successful, the court will generally order the other party to stop using the mark, and the court will sometimes require the defendant to pay money to the owner of the mark to compensate for economic harm caused by the customer confusion resulting from the unauthorized use.

The owner of the mark must also be able to prove that the defendant used the mark (or one substantially similar to it) in a manner resulting in confusion to the customers of the owner. Trademark law essentially provides a legal mechanism to permit organizations that invested resources into the creation of a commercially recognized mark to protect that investment. Thus, one of the key factors necessary to support a finding of misappropriation is proof that the unauthorized use of the mark actually confused current or potential customers of the mark owner. The more distinctive the mark, the stronger the claim of trademark ownership. The strongest trademark ownership claims are associated with marks that have no meaning other than that associated with the company that created it and with the products of that company.

Marks that have "generic" meanings (i.e., meanings that have nothing to do with the company in question) are more difficult to protect as trademarks. Marks that have generic meanings can only be established as trademarks to the extent that a "secondary meaning" is established by the commercial user. The secondary meaning is the commercial meaning that the consuming public recognizes and associates with a specific company and its products. Two examples of trademarks that have both generic and trademark status are "Playboy" and "Apple." Both of those words have generic meanings, but each has also been established as an enforceable trademark through commercial use and the development of a secondary meaning associated with a specific company. When a mark has both a generic meaning and a secondary meaning as a trademark, the issue of whether a particular use of the mark by another party is trademark misappropriation or not is a question of fact that depends on the actual circumstances of the unauthorized use. In such cases, the law protects only the investment applied by the mark owner to establish the secondary meaning for the mark. The law does not permit the trademark owner to block uses of the mark that involve the mark's generic meaning.

Accordingly, Apple Computer can effectively enforce its trademark rights as to the mark, "Apple," in its classes of use (e.g., the computer industry), but it cannot stop parties who use that mark in other fields in which the term is generic and thus not protected as a trademark (e.g., the produce industry).

Note that distinctive marks that have strong trademark protection can deteriorate into generic terms that are not protected by trademark law if the owners of the mark do not effectively manage the use of the mark. Terms such as "aspirin" are now widely used as generic words describing a certain type of product. Initially, however, those marks were developed and protected as trademarks, based on their secondary meaning associated with one company and its products. Over time, however, the name became publicly associated with an entire class of products manufactured by several different companies. Once that condition developed, the mark lost its trademark status and became a generic word that can be used commercially by anyone. Trademark owners must challenge uses of their marks that link those marks to products other than those of the owner of the trademark. Failure to stop such generic use of the trademark will lead to a situation where the mark has lost its value as a commercial identifier for one company and its products. At that point, the mark is no longer considered an enforceable trademark and it is not subject to trademark law protection.

In the United States, there is another legal right possessed by trademark owners. This additional right is the right to prevent "dilution" of the trademark. Trademark dilution occurs when an unauthorized user of a trademark (or one that is similar to the mark in question) makes use of the mark in a way that reduces the economic value of the mark. The primary remedy available to a trademark owner if dilution can be proven is economic compensation for the reduced value of the mark resulting from the unauthorized use. The trademark dilution claim differs from the basic trademark misappropriation claim since it does not focus on customer confusion. The dilution claim, instead, compensates a mark owner for any reduction in the economic value of the mark that was caused by the misuse of the mark (or one similar to the claimed mark). In a sense, the dilution claim is more of a legal protection for property value associated with a mark, while the traditional trademark claim provides protection for competitive harm resulting from customer confusion.

To raise a claim of dilution successfully, a trademark owner must first demonstrate that its mark is "famous." Several basic factors regarding a mark are considered by courts as they determine whether or not the mark in question is famous. Courts look to see whether the mark is distinctive: A more distinctive mark is more likely to be viewed as famous. Another factor considered is the duration and extent to which the mark has been used with the specific goods in question; the longer and more extensively such use has been made, the more likely a court will find the mark to be famous. The duration and extent of the advertising and publicity afforded a mark are also considered. The greater the investment in such promotion for the mark, the more likely the mark will be seen to be famous.

Courts also examine the geographic extent of the trading are in which the mark is used in commerce, with a greater range more likely to support a claim that the mark is famous. Courts also examine the channels of trade through which the mark is distributed in commerce: The more prominent the channel, the more likely there will be a finding that the mark is famous. The extent to which the mark is recognized in the trading area and the channels of trade is also an important factor, and if the recognition is high, the mark is more likely to be treated as a famous mark. Also examined are the nature and extent of use of the same or similar marks by other parties, with such other use undermining a finding that the mark is famous. Finally, courts look to see if the mark has been registered with the government, and, if it has been registered, courts are more likely to find the mark to be famous.

When a trademark has been properly registered with a government authority, it should be accompanied, when published, with the appropriate trademark registration sign (®). When a party plans to assert trademark rights through common law use of the mark instead of registration, the mark should be followed by "TM" when it is published. Use of the appropriate indication of either registration or common law usage is important, as it puts other parties on notice that the mark is claimed as a trademark. If the appropriate trademark notice is used with the mark, a party who misuses the mark can be subject to additional penalties as a "willful infringer." If the proper notice is not provided, a party who misuses the mark may be classified as an "innocent infringer" and thus receive a less rigorous penalty than that applied to a willful infringer.

In most jurisdictions, trademarks do not have a specific term. They do not generally expire. Instead, they remain enforceable for as long as the owner continues to use them in commerce. When a trademark owner ceases to use a mark commercially, that mark is deemed to be "abandoned." Abandoned marks can then be used by other parties, without liability.

4.2 Trademarks and the domain name registration process

Trademark rights are substantially connected with the Internet domain name registration process. The significant commercial demand for domain names has led many enterprises to try to establish their brand identity on the Internet. Domain names became the most popular way to accomplish this competitive activity. Domain names raise trademark challenges in at least two important ways. First, there is the issue of more than one party attempting to integrate the same trademark into a registered domain name. Second, there is the challenge of asserting trademark rights over a domain name itself.

The domain name registration process has developed in a manner that permits the grant of domain names without prior trademark review. Domain name registrars do not, as a matter of course, conduct trademark searches prior to registering a new domain name. This approach was partially driven by the fact that, given the large number of domains being registered, effective trademark searches for all applications would place a large administrative burden on the registrars.

Additionally, registrars were concerned about potential legal liability if they conducted such searches but inadvertently registered domains to parties other than the rightful owners of the trademarks incorporated into those domains. Registrars feared that if they undertook the role of reviewing proposed domain names to see if those names infringed on trademarks of other parties, they could be sued as "contributory infringers" of a mark, if they made an error and registered a domain name that did infringe on trademark rights of others. The registrars generally took the position that they would face less potential liability if they implemented registration procedures that offered no implication that they were monitoring registrations to reduce incidents of unauthorized trademark use in domain names.

Domain name registrars have also been challenged by would-be registrants who have had proposed domain names rejected by the registrars for policy reasons. For example, as a matter of course, many registrars refuse to register domain names that are offensive or have sexual meanings. Rejected registrants have challenged those actions by the registrars, claiming that such rejections are unconstitutional. To date, however, the U.S. federal courts have rejected those arguments, noting that constitutional law protections such as the First Amendment right of free expression are applicable only to actions by governments, and the domain name regis-trars are not viewed by the courts as parties acting under the color of gov-ernment authority (*National A-1 Advertising, et al. v. Network Solutions, Inc.* and *Island Online Corp. v. Network Solutions, Inc.*).

In addition to incorporating trademarks into domain names, domain names themselves can also be protected as trademarks. That protection does not, however, arise immediately upon registration. Registration of a domain name alone does not grant the party who registered the domain trademark rights. In the United States, for example, a federal court, in the case of *Brookfield v. West Coast*, concluded that domain name registration must be coupled with actual commercial use of a domain name in order to establish trademark rights in a domain name. To establish trademark rights over an entire domain name, the party must follow the same requirements for creating an enforceable trademark that it would follow for any other form of trademark (e.g., a phrase, a logo). Specifically, the party seeking trademark status for the domain name must obtain trade-mark registration status from a government or it must develop common law ownership rights through commercial use of the mark. If a party wants to create trademark rights for an entire domain name, then it must obtain government registration for the trademark or actively use the domain name in commerce and effectively document that use.

The U.S. Patent & Trademark Office, supported by federal courts, has concluded, however, that the nondistinctive elements of Universal Resource Locators (URLs) are not subject, on their own, to trademark protection. For example, the PTO and a federal court concluded that com-mon URL modifiers such as "http" and "www" are not subject to trade-mark law protection, independent of distinctive elements of a domain name (*Image Online Design v. Core Associates*).

Trademark law is actively used by mark owners to stop use of their marks by other parties in domain names. Major businesses that have invested substantial resources in the creation of their trademarks move aggressively to apply trademark infringement and dilution principles to block other parties who attempt to use identical or similar marks in domain names (*Hasbro, Inc. v. Clue Computing, Inc.* and *Porsche Cars North America v. Porsch.com*). Trademark owners have also tried to use trademark law theories to stop operators of sites that express criticism of the mark owners (e.g., *Bally Total Fitness Holding Co. v. Faber*); however, those efforts have not been particularly successful.

4.3 The problem of cybersquatting

Cybersquatting is the process through which a party, other than the owner of a trademark, seeks to register that mark as part of a domain name, with the intention of deriving economic advantage from the use of the mark in the domain name. Cybersquatting generally involves trademark misappropriation and trademark dilution; thus, in the early days of commercial Internet use, cybersquatters were commonly challenged through use of trademark lawsuits. As with all trademark misappropriation cases, plaintiffs in trademark cases involving domain names registered by other parties must demonstrate that they have a superior claim to the trademark and that the domain name registration granted to the other party constitutes an infringement on that mark. In these cases, generally, the plaintiff must prove that it has established enforceable trademark rights and that the use of the mark in the domain name by the other party will likely lead to customer confusion. If a dilution claim is raised, the plaintiff must demonstrate that the use of the mark in the domain name reduces the economic value of the mark. In those trademark actions, the defendant can prevail if it can demonstrate that it has a superior right of use for the mark or that its use of the mark in the domain name is not likely to result in customer confusion.

Trademark owners now have an additional legal claim that they can raise in some instances when their trademark has been incorporated into a domain name registered by another party. This additional claim is known as "cybersquatting." Cybersquatting is now prohibited by the rules established for the domain name registration process by the Internet Corporation for Assigned Names and Numbers (ICANN) and by federal law in the United States (the Anticybersquatting Consumer Protection Act of 1999 [ACPA]). Thus, today, trademark owners often pursue simultaneously several different legal claims against cybersquatters. It is common to see trademark misappropriation and dilution claims raised along with claims under the anticybersquatting statute, as well as private arbitration actions brought under the ICANN dispute resolution process.

ICANN is a private, nonprofit, international corporation that plays a leading role in the management of the domain name registration process. The actual legal limits of ICANN's authority remain uncertain, at present, since ICANN operates based on a somewhat ambiguous heritage. ICANN is not a government entity, and it was not formally established by any government entity. ICANN has contractual relationships with the U.S. government and many private parties (including all of the authorized domain name registrars); however, ICANN's actions do not carry the force of law in the United States or any other jurisdiction. ICANN's domain name rules and processes function as all of the key players involved in Internet operations (e.g., governments, service providers, users) continue to be willing to abide by those rules and processes. In the future, however, there will likely be more disputes and challenges to ICANN's authority. But for the present, ICANN continues to be the key player in the global domain name registration process.

The ICANN rules regarding cybersquatting and the ACPA provisions are substantially similar as to the definition of cybersquatting; however, there are some subtle differences. ICANN requires that a party raising a domain name complaint (the complainant) must prove the following elements against the other party (the respondent) in order to win the complaint: that the domain name registered by the respondent is identical (or confusingly similar) to a trademark in which the complainant has enforceable legal rights, that the respondent has no legitimate rights of use for the domain name, and that the respondent registered the domain name in bad faith. ICANN defines bad faith as any one of the following motivations: a desire to prevent the owner of the trademark from using the mark in a domain name, an intent to sell (or transfer in any way) ownership of the domain name for economic gain greater than the cost of registration, an effort to disrupt the business activities of a competitor, or an intent to derive economic gain by attracting users to a Web site.

Under the ACPA, a party contesting a domain name registration (the plaintiff) sues the party who registered the domain name (the registrant or the defendant) in federal court. To win the case, the plaintiff in an ACPA action must persuade the court of the following elements: that the plaintiff has an enforceable ownership interest in the mark, that the defendant acted in bad faith when it registered the domain name, that the domain name is identical or confusingly similar to the plaintiff's mark, the domain name dilutes the mark, or the domain name is a mark that is protected under some specific law. Bad faith, under the ACPA, is defined as any one of the following elements: providing false or misleading information, intent to profit from registration of the mark, intent to sell or otherwise transfer rights of use in domain name for economic gain, intent to divert traffic from the mark owner's Web site, "warehousing" of domain names, or registration even though the defendant is aware it has no valid intellectual property law rights to the mark.

Defenses to ICANN and ACPA domain name claims also vary somewhat. Under the ICANN rules, a respondent can defend itself by demonstrating that it had actually used the domain name in its commercial operations prior to receipt of notice of the claim. Another defense is that the respondent's use of the domain name constitutes noncommercial, fair use of the mark. Finally, the respondent can argue that it is publicly known by the mark and is thus entitled to use the mark in the domain name. Note that these defenses are, in effect, challenges to the trademark ownership rights asserted by the complainant.

Some arbitrators applying the ICANN rules also recognize a counterclaim raised by respondents against complainants. This counterclaim has been described as a claim of "reverse domain name hijacking." Respondents who can demonstrate that the complainant raised the domain name complaint in spite of the fact that the complainant was aware that the respondent had a legitimate interest in the domain name, or the complainant was aware that there was an absence of bad faith on the part of the respondent, can win a reverse domain name hijacking counter-claim. This claim is, essentially, an abuse of process argument raised by a respondent against a bad faith complainant. Examples of UDRP arbitration cases in which this theory was applied include *Goldline International v. GoldLine*; *Smart Design v. Carolyn Hughes*; *Koninklijke v. Telepathy, Inc.*; and *Deutsche Welke*. It is unclear, however, what remedies are available to respondents in these cases, since presumably they would retain ownership of the registered domain name even without the counterclaim when the complainant was unable to prove the required elements to sustain its complaint.

The ACPA provides somewhat different defenses for a domain name defendant. One ACPA defense is prior use of the domain name. This defense requires a showing that the defendant made bona fide commercial use of the domain name prior to filing of the claim by the plaintiff. Another ACPA defense is that the domain name is the legal name of the defendant. These defenses also consist, in part, of challenges to the underlying trademark rights of the plaintiff in the ACPA action. Another available defense is a fair use argument, based on a claim that the registrant registered the domain name for a permissible purpose. An example of such permissible fair use is registration of a domain name for a Web site presenting legitimate, good faith criticism of the trademark owner (*Northland Insurance Co. v. Blaylock*).

Cybersquatting can only be raised effectively as a claim if the party raising the claim has a superior right to the trademark in question than the party who registered that mark in a domain name. In addition, cybersquatting requires some level of bad faith on the part of the defendant. A good faith action does not constitute cybersquatting. The party registering the mark must have some intent to profit or to harm the mark owner before a cybersquatting claim can be sustained. The cybersquatting claim cannot be used against a party who has a legitimate right to use the mark in question or one who lacks the requisite bad faith intent either to profit from the use or to injure the owner of the mark as a result of the use.

Both the ICANN and ACPA provisions can be applied against registered domain names that contain slightly misspelled versions of trademarks. This practice is commonly described as "typosquatting." Typosquatting is prohibited by anticybersquatting rules, provided that the basic elements of those rules are met. Accordingly, cases involving typosquatting also require a factual showing that there is bad faith on the part of the registrant and a superior claim on the part of the plaintiff.

4.4 Resolving domain name disputes

The ICANN anticybersquatting rules are enforced through the Uniform Dispute Resolution Procedures (UDRP), which provide for arbitration to resolve cybersquatting claims. UDRP arbitration is handled by the United Nations World Intellectual Property Organization (WIPO) and by several private dispute resolution organizations. The U.S. statute is enforced by the federal courts in the United States. Both systems require that the private party alleging ownership of the mark incorporated into a domain name raise the claim directly against the party who registered or who seeks registration of the domain name.

Remedies available for cybersquatting differ between the UDRP and the U.S. law. The UDRP process provides only for cancellation of domain name registrations received by a cybersquatter and for transfer of the registration to the legitimate owner of the mark. The anticybersquatting law in the United States provides for monetary damage awards, to be paid by the cybersquatter to the owner of the mark as compensation for the misuse of the mark. The U.S. law also authorizes courts to issue orders requiring the cybersquatter to transfer the domain name registration to the trademark owner.

The UDRP arbitration process has become popular, primarily for two reasons. First, it is generally a faster and less expensive process than traditional litigation. Most of the process can be handled electronically, and the schedule for processing of claims is compressed. The process is particularly popular when the parties are located in different countries. The second reason for its popularity is that the UDRP forum has been a highly successful one for trademark owners. A very high percentage (reportedly more than 80%) of trademark owners who file UDRP claims win those claims.

Arbitration applying the rules established by ICANN through the UDRP is conducted by several organizations. One of the leading UDRP domain name arbitration systems is operated by the World Intellectual Property Organization of the United Nations. Currently, the WIPO arbitration system handles the bulk of domain name arbitration cases (www.arbiter.wipo.int/domains). In addition to the WIPO system, however, domain name dispute arbitration cases are handled by private groups, including the following: the National Arbitration Forum (www.arbforum.com), the CPR Institute for Dispute Resolution (www.cpradr.org), and eResolution (www.eresolution.ca).

An open issue associated with these different domain name dispute resolution rules and procedures is the question of how judgments set in one forum will be enforced or respected by others. For example, a federal court of appeals in the United States reversed a federal district court and granted an injunction that reversed an arbitration opinion issued under the UDRP (*Sallen v. Corinthians Licenciamentos, Ltd.*). In that case, a UDRP arbitration action resulted in an order that Sallen surrender the domain name he had registered. Sallen went to federal district court in the United States asking for a court order to reverse the UDRP ruling. The federal district court refused to grant the injunction, determining that it did not have jurisdiction over the UDRP judgment. The First Circuit federal court of appeals reversed the district court judgment and granted the injunction in favor of Sallen.

The *Sallen* case illustrates the fact that there will be a need to reconcile decisions made by UDRP arbitrators and those made by courts in countries around the world. Expect more examples of conflicting judgments regarding domain name ownership in the future. Remember that these conflicts between arbitrators and courts can result in greater uncertainty as to domain name ownership. Many different countries are currently attempting to determine how much deference they will apply to judgments of UDRP arbitrators. A federal court in the United States, for example, determined that decisions made by UDRP arbitrators would not necessarily receive the same official status as arbitration rulings sanctioned by the U.S. Federal Arbitration Act (*Parisi v. NetLearning, Inc.*).

Note that the ACPA provides for court jurisdiction in the judicial district in which the relevant domain name registry is located. In the U.S. case, *Mattel v. barbie-club.com*, the federal court concluded that a lawsuit brought under the ACPA could be brought in the court located in the same judicial district in which the registry that manages the domain name in question is physically located. This means that parties who registered domain names using a registry located in the United States can be subject to lawsuits in the U.S. courts in the event of a claim under the ACPA, even if the party who registered the domain has no other presence in, or contact with, the United States (*Cable News Network v. CNNews.com*).

Courts in a variety of countries now routinely handle cases involving disputes arising from domain name registration and use. A Czech court, for instance, issued an injunction barring use of a registered domain name pending resolution of the dispute over an associated trademark (*Quilt SRO* Case). Another example is provided by a French court's determination that domain name registrars are not liable for trademark law claims arising from names that they registered (*Agence Nationale Pour L'Emploi v. 7 Ways*).

4.5 Reconciling the interests of legitimate trademark owners

Efforts to curb cybersquatting highlight the challenge of reconciling competing demands for the same domain name by different legitimate owners of a single trademark. Under trademark law, it is common for various organizations to possess ownership rights for the same mark or for ones that are substantially similar. This happens, since enterprises that do not compete for the same customers have the right to make use of the same mark. Trademark rights were intended to protect businesses from activities that might confuse customers; however, if two businesses are not competing for the same customers, use of the same mark by both parties would not cause a trademark conflict. Commonly, companies working in different industries or in different geographic regions can simultaneously use the same trademark with no adverse consequences. Trademark law thus permits multiple owners of a mark, provided that there is little likelihood of customer confusion.

Domain name registrars have traditionally not taken the issue of multiple legitimate trademark owners into account. The "first come first served" approach to domain name registration led to a system in which the first party to register a domain based on its trademark was effectively able to preempt all other owners of that mark by making it virtually impossible for them to complete a competing registration. This has led to a situation where trademark owners now commonly register essentially all of their trademarks in domain names, along with typographical variations on those marks and negative or critical versions of their marks (e.g., "xyz-corpstinks.com"). This widespread registration is, in part, a defensive action directed toward protecting the trademark owner from both cybersquatters and other legitimate trademark owners.

One approach initiated to try to address the problem of multiple legitimate trademark owners was ICANN's creation of additional top-level domains (TLDs). Those additional TLDs are .biz, .info, .pro, .name, .aero, .museum, and .coop. The concept behind the expansion of TLDs was that the additional TLDs would make it possible for more parties to register the same domain name. The same domain name can now be registered in more than one TLD, thus, in theory, permitting more than one legitimate mark owner to be able to use the same mark in domain names. In practice, however, this initiative has not been as helpful as hoped, primarily for two reasons. The first is that the number of new TLDs was not large enough to permit accommodation of many other mark owners. The other problem is caused by the fact that registrants now simply register their marks in multiple TLDs as a matter of course. In effect, this strategy results in extension of the first come first served registration process into multiple TLDs.

Another part of the effort to increase the number of trademark owners who can use the same mark in their domain names is the rise of alternative domain name registrars (e.g., www.new.net). Although they do not control formally recognized TLDs, these systems simulate additional TLDs through use of special software housed on certain servers. This process makes it appear to an end user that there are additional TLDs (e.g., .kids, .travel), and it can enable more parties to incorporate the same mark into their domains. However, because these simulated TLDs have not been recognized by ICANN, they are not universally accessible, and they are not true TLDs. Although these alternative domain name systems have the potential to ease the competition among trademark owners for scarce domain names, some mark owners are concerned that this process can lead to additional trademark law conflicts, and they oppose the systems. This controversy has led some of the alternative domain name service operators to ask a federal court for a declaratory ruling indicating that the systems do not violate trademark law requirements (*Neulevel, Inc. v. Amazon.com, Inc.*).

Widespread commercial use of the Internet makes it difficult for multiple trademark owners to use the same mark online without causing some commercial confusion. While in the past it was relatively easy for different businesses operating in different commercial markets to use the same mark simultaneously and avoid customer confusion, that coexistence is now much more difficult when the multiple owners of the mark all want to use the mark online. In an increasing number of instances, the mere fact that two legitimate trademark owners both want to make use of their mark for online commerce may result in an irreconcilable trademark conflict, even though both parties had previously shared the mark with no consumer confusion. For example, if a company in Australia has established trademark rights in that country for a particular mark, and a different company in England has done the same for that mark in that country and in a different commercial industry, those two companies would traditionally be able to coexist, each using the same mark in its different industry and different geographic region. If, however, both try to register the mark as part of their domain names in the same TLD, only one will be able to obtain a registration. As a consequence, one legitimate trademark owner will be unable to use its mark in its domain name. At present, no truly effective resolution to this problem has been implemented.

4.6 Domain name property rights

In addition to trademark and anticybersquatting laws, some courts are now willing to consider application of traditional property law rights to domain names. Under this legal theory, domain names could be treated in the same manner as personal property. Very few jurisdictions have addressed this theory, and among those that have, the results are conflicting. For example, a trial court in Virginia accepted the basic premise that a registered domain name could be protected under personal property law; however, that theory was rejected by the Virginia Supreme Court in an appeal of the lower court decision (*Network Solutions v. Umbro*). In another case, however, a federal court in the United States held open the possibility that domain names could be subject to traditional property law rights. Both of these U.S. cases examined the property rights issue in the context of efforts by creditors to place liens on the registered domain names of parties in default (*Dorer v. Arrel*).

In its rejection of the property law theory for domain names, the Virginia Supreme Court concluded that domain names were creations of service contracts between domain name registrars and the owners of the domains. That court thus took the position that rights of control over domain names were actually elements of service contracts, not rights linked to a new form of personal property. Other courts have also taken this position, treating domain name ownership rights as contract rights instead of property ownership rights (*Zurakov v. Register.com*). The scope of applicability of the property law argument for domain names is currently far from settled.

4.7 Country code TLDs

Top-level domains are divided into two categories. One category consists of the generic TLDs (e.g., .com). The other category consists of country code TLDs (e.g., .uk). Registration for generic TLDs (gTLDs) is controlled by private organizations. Registration for country code TLDs (ccTLDs) is controlled by governments, although some governments now contract with private entities to manage their domains. There are approximately 244 ccTLDs, and of those TLDs, approximately 195 are actively registering domains. Certain ccTLDs are particularly attractive, from a commercial perspective. Tuvalu's ".tv" domain and Turkmenistan's ".tm" are two examples of ccTLDs that are highly prized by some commercial enterprises.

Increasingly, domain name disputes are arising in the ccTLD context, as diverse parties attempt to register similar marks as domain names in the ccTLDs. Increased global commercial use of the Internet makes businesses eager to register domain names in multiple ccTLDs. It is now a common practice for companies to register their domain names in several ccTLDs, in addition to the basic generic TLDs. This increased demand for domains in ccTLDs also leads to disputes as multiple applicants attempt to register names. The provisions of trademark law generally apply to ccTLD domain name registration; however, the trademark law applied would most likely be the local trademark law of the country controlling the ccTLD in question. It is important to note that many of the ccTLDs do not apply the

ICANN UDRP regulations, and some of those ccTLDs do not have clear dispute resolution policies or procedures. Trademark owners attempting to protect their marks from unauthorized use in foreign ccTLDs face a greater challenge than that associated with protection of their marks in the various generic TLDs (e.g., .com), since they do not have a convenient forum for those actions and they do not have a uniform set of rules to apply.

Several highly visible domain name disputes involving ccTLD registrations illustrate the difficulty in managing domain names in multiple TLDs. It has been reported that America Online was forced to litigate in Brazil for control of "aol.com.br" and Kodak engaged in arbitration over "kodak.ru" in Russia. Similar conflicts were reportedly faced by Amazon.com in Greece and Playboy Enterprises in Italy. Expect the number of these ccTLD domain name disputes to continue to increase for the foreseeable future.

The registration process for the diverse ccTLDs varies somewhat from country to country. Approximately 96 of the ccTLDs are described as "unrestricted" ccTLDs. This means that there is no local presence in the country required as a condition for registration of a name in the ccTLD. Approximately 99 of the ccTLDs are characterized as "restricted." The restricted ccTLDs require some form of local presence (either a physical presence in the country or a company formed under the laws of that country) prior to domain registration. Examples of unrestricted ccTLD countries include Venezuela, Belgium, the United Kingdom, Denmark, the Philippines, Mexico, and the Czech Republic. Examples of restricted ccTLD countries include Germany, France, and Japan.

As more parties register names in diverse ccTLDs, the specific local restrictions associated with those registrations become a greater challenge. For example, some countries take the position that registration of a domain name in the country's ccTLD is sufficient to establish legal jurisdiction over the registrant, even if the registrant has no physical presence in the country. Once legal jurisdiction is established, the registrant can be subject to the laws of the country and can be brought into litigation in that country. Accordingly, it is possible that a U.S. company, with no presence in Belgium, for example, could become subject to legal jurisdiction in Belgium after receiving domain name registration in the ccTLD controlled by Belgium. That jurisdiction could be asserted through enforcement of the laws of Belgium (e.g., tax or commercial transaction laws) against the company or through court jurisdiction enabling parties with grievances against the company to litigate those claims in courts in Belgium. The potential legal consequences of ccTLD registration should, therefore, be considered prior to registration.

4.8 Multilingual domain names

Increasing international use of the Internet is leading to demand for multilingual online content. One important aspect of that demand is a movement toward domain names registered in languages other than English. Regardless of the language used in a domain name, the anticybersquatting and trademark laws continue to apply. Thus, one can not avoid a claim of cybersquatting or of a trademark violation simply because the registered domain name is a translated version of a protected mark. For example, in *Sankyo Co. v. Zhu Jia Jun*, an arbitration action brought under the UDRP, arbitrators found in favor of the trademark owner even though the registered domain name at issue was a Chinese language version of an English language mark. A trademark in one language can be protected by its owner against registration by another party as part of a domain name in another language.

4.9 Trademarks and hypertext links

Links between pages of different Web sites also present the potential for trademark disputes. When names or logos that are claimed as trademarks are used as the icons for hypertext links between Web pages, there is the potential for trademark law controversy. Generally, such use of the marks does not cause a legal issue if the link has been authorized by the owner of the trademark and if the context in which the mark is presented in the linking site is not offensive to the mark owner. Caution is advisable, however, when using a name or other symbol that is a trademark of another party for the purpose of linking Web content. The safest approach is to obtain prior consent to use the mark as the link icon. An alternative to express permission is a statement in the notices and disclaimer section of your site indicating that the site may contain material claimed by other parties as trademarks and that use of those marks in the site is incidental fair use and does not constitute any assertion of ownership rights in those marks on the part of the site operator.

4.10 Keywords and trademarks

Trademark law issues also arose in the context of online keyword search systems. For example, trademark disputes arose when some Web site operators began to embed keywords that were claimed as trademarks by other parties in the metatags used to attract search engines. Another trademark issue developed when search engine operators began to sell on-screen advertising associated with keywords used by parties who are conducting online content searches. There is a trademark law issue if the keywords that are "sold" for advertising are trademarks claimed by other parties.

When metatags include words claimed as trademarks by other parties, courts have been willing to find trademark misappropriation. In the United States, several cases illustrate the fact that unauthorized use of the trademarks of another party in metatag key words can result in liability under trademark law (*Playboy Enterprises v. Calvin Designer Label*). Courts in the United States are particularly likely to find liability when the parties involved are direct competitors (*Insituform Technologies v. National Envirotech Group*). Courts in the United States have not, however, applied the ACPA against use of another party's trademark in metatags (*Bihari v. Gross*).

Trademarks used by parties other than their owners can also raise disputes in the context of keyword advertising. Search engines commonly sell advertising rights to keyword search terms, to help advertisers target their ads more effectively. When the keywords purchased for online advertising purposes are trademarks owned by other parties, there is a basis for a trademark claim. Also note that courts outside of the United States now address the trademark and keyword issues. For example, an Italian court concluded that unauthorized use of another party's trademark as a metatag keyword constituted unfair competition (*La Triests e Venezia Assicurazioni Genertel v. Crowe Italia*).

The standard practice is to avoid using marks that are claimed as trademarks as keywords for either metatag or advertising purposes. Trademarks that also have standard generic meanings can be used as keywords to the extent that the meaning that is intended by the user is the generic meaning and that the usage does not result in customer confusion (*Playboy v. Netscape*). Special attention should be paid to keyword selection when those keywords bear any type of relationship with competitors.

4.11 Unfair competition claims

Trademark and anticybersquatting law claims are commonly supplemented by legal claims based on the legal theory of unfair competition. Unfair competition is a legal claim that is distinct from trademark law but is often used in conjunction with trademark law arguments. Most countries prohibit commercial practices that are unfair to competitors or customers. Precisely what constitutes unfair competition varies substantially among jurisdictions.

It is common for unfair competition law claims to be raised in disputes involving domain names and other forms of commercial online content. For example, in addition to trademark and anticybersquatting claims, an owner of a trademark is likely to argue that registration, by one of its competitors, of a domain name incorporating the mark is a form of unfair competition, since it would unfairly undermine the owner's ability to compete. That claim could be based on the fact that the registration would prevent the owner from establishing a domain name using its mark, or the claim could be based on an argument that the registrant plans to use the domain name to confuse potential customers. Similarly, a trademark owner could claim that use of one of its marks as an embedded metatag keyword by a competitor will result in confusion to potential customers, thus causing harm to the competitive environment.

Unfair competition law thus provides a flexible legal basis for claims by trademark owners who believe that their marks have been misused in an online commercial context. It can be used when the conduct in question may not qualify for relief under trademark law or anticybersquatting regulations, but the impact of the conduct is particularly unreasonable from a commercial competition perspective. This flexibility is the primary reason why competition law claims are increasingly popular with regard to domain names and other online content.

4.12 Strategies to manage domain names and keywords

Trademark owners should apply basic strategies to manage their marks in the context of domain names and online keywords. With regard to domain names, trademark owners should recognize that they are engaged in a race against others for the right to use their marks in domain names. In that environment, they should be prepared to register the domain names based upon their marks in multiple TLDs and in different languages (e.g., multilingual domain names). After registering those domains, they should establish a process through which they can be sure that all actions necessary to maintain their rights to the domains (e.g., registration renewals) are conducted as required. They should monitor domain name registrations to make sure that their trademarks are protected from both cybersquatting, typosquatting, and conflicting use by others who may have some rights of use to those marks. When conflicting registrations are discovered, the mark owner should make use of all available mechanisms to defend its ownership of the mark in question (e.g., UDRP arbitration, anticybersquatting litigation, trademark or unfair competition law claims). Recognize that this defense may require litigation or arbitration in foreign countries when the registration is in a foreign ccTLD.

When asserting trademark ownership, it is important that the owner of the mark enforce its claim against all other users. Failure to assert trademark rights over parties who are using the same mark, or one substantially similar to it, can undermine the strength of the mark owner's claim. If a mark owner is aware of unauthorized use but fails to take action to stop that use, that failure can be cited by other parties as an indication of abandonment of the mark. It is thus essential that trademark owners monitor commercial activities in their industry (i.e., class of use) to be on the lookout for misuse of their marks. When possible instances of such misuse are found, the mark owner must investigate to see if there is actual infringement or misappropriation, and, if there is, the mark owner must take all reasonable steps to stop the misuse of the mark. In order to maintain enforceable trademark rights, a trademark owner must police its marks and take action to stop unauthorized use or run the risk that its trademark claims will be eroded.

Also remember that in the United States, the trademark dilution claim permits a mark owner to take legal action even in instances when there is no clear customer confusion resulting from the misuse of the mark. Dilution claims are increasingly popular in the context of domain name and other online content disputes (*Avon Products v. Lew*, *Hasbro v. Internet Entertainment Group*, and *Toys R Us v. Mohamed Akkaoui*). Note, however, that in order to recover compensation for a dilution claim, the owner of the mark must quantify the economic value of the mark in question and quantify the resulting loss in value of the market resulting from the unauthorized use (*Avery Dennison v. Sumpton*). If the owner is unable to establish the value of the mark, a dilution claim will not be sustained.

Owners of marks who intend to assert trademark rights for those marks should be sure to use those marks in exactly the form that they want to protect. Variations in a mark, as used in commerce, can undermine the ability of the mark owner to apply the protection afforded by trademark law. In addition, registration and common law usage notices should be properly displayed with each claimed trademark. Online notices should clearly indicate all trademarks that are claimed by the content provider.

Parties using marks other than their own trademarks should also adopt basic management strategies for online use of those marks. These strategies should be applied to help reduce the risk of legal liability. Whenever possible, avoid using trademarks of other parties in any online setting, unless prior approval from the trademark owner is obtained. That prior approval should be obtained in written form, and the document should be retained in an accessible location. Searches of registered trademarks and marks claimed through common law usage should be conducted before prominent online use of nongeneric words. Although these searches are not foolproof, it is a good idea to conduct them prior to using a mark as part of a domain name or for online keyword purposes. By conducting such prior searches, it may be possible to avoid liability as a willful infringer if there is a future claim of infringement. If you intend to use a word that has both a generic meaning and is claimed as a trademark, make sure that your use of the word is not likely to result in customer confusion and that it is based on the generic meaning of the word, not its trademark usage. Avoid using the word if it is claimed as a trademark by one of your competitors or customers.

Selected bibliography

Agence Nationale Pour L'Emploi v. 7 Ways, Tribunale de Grande Instance de Nanterre (France 2001).

Avery Dennison Corp. v. Sumpton, 99 F. Supp. 1337 (C.D. Cal. 1998).

Avon Products, Inc. v. Lew, Case No. 99 Civ. 1213 (S.D. N.Y. 1996).

Bally Total Fitness Holding Co. v. Faber, 29 F.Supp.2d 1161 (C.D.Cal. 1998).

Bihari v. Gross, 2000 U.S. Dist. LEXIS 14180 (S.D. N.Y. 2000).

Brookfield v. West Coast, Case No. 98-56918 (9th Cir. 1999).

Cable News Network, L.L.P. v. CNNews.com, 162 F. Supp. 484 (E.D. Va. 2001).

CPR Institute for Dispute Resolution, <www.cpradr.org>.

Deutsche Welke, Case No. D2000-1202 (2000).

Dorer v. Arrel, 60 F. Supp. 2d 558 (E.D.Va. 2000).

eResolution, <www.eresolution.ca>.

FleetBoston Financial Corp. v. fleetbostonfinancial.com, 138 F. Supp. 2d 121 (D. Mass. 2001).

Goldline, International, Inc. v. GoldLine, Case No. D2000-1151 (2000).

Hasbro, Inc. v. Clue Computing, Inc., 1999 WL 711429 (Sept. 2, 1999).

Hasbro, Inc. v. Internet Entertainment Group, Ltd., Case No. C96-130WD (W.D. Wash. 1996).

ICANN Rules for Domain Name Dispute Resolution, at <www.icann.org/udrp/udrp-rules-oct99.htm>.

ICANN Uniform Dispute Resolution Policy, at <www.icann.org/udrp/udrp-policy-24oct.htm>.

Image Online Design v. Core Assoc., et al., 2000 U.S. Dist. Lexis 10259 (C.D. Cal. Jul. 21, 2000).

Insituform Technologies, Inc. v. National Envirotech Group, LLC, Case No. 97-2064 (E.D. La. 1997).

Island Online Corp. v. Network Solutions, Inc., Case No. 99-CV-6848 (DGT) (S.D.N.Y. Nov. 13, 2000).

Koninklijke KPN NV v. Telepathy, Inc., Case No. D2000-0217 (2001).

La Trieste e Venezia Assicurazioni Genertel v. Crowe Italia, Court of Rome (Italy 2001).

Mattel, Inc. v. barbie-club.com, 2001 U.S. Dist. LEXIS 5262 (S.D. N.Y. 2001).

National A-1 Advertising, et al. v. Network Solutions, Inc., et al., Civ. No. 99–033-M (D. N.H. Sept. 28, 2000).

National Arbitration Forum, <www.arbforum.com>.

Network Solutions, Inc. v. Umbro, International, Inc., 529 S.E. 2d 80 (Va. 2000).

Neulevel, Inc. v. Amazon.com, Inc., Case No. 011245 (E.D. Va. Sept. 6, 2001).

Northland Insurance Co. v. Blaylock, 2000 U.S. Dist. LEXIS 14333 (D. Minn. 2000).

Oppendahl & Larson v. Advanced Concepts, Case No. 97-3-1592 (C.D. Col. 1997).

Paine Webber, Inc. v. www.painewebber.com, Case No. 99-0456-A, 1999 U.S. Dist. LEXIS 6552 (E.D. Va. 1999).

Parisi v. NetLearning, Inc., 139 F. Supp. 2d 745 (E.D. Va. 2001).

Playboy Enterprises, Inc. v. Calvin Designer Label, 44 U.S.P.Q. 2d 1156 (N.D. Cal. 1997).

Playboy Enterprises, Inc. v. Netscape Communications, Corp., 55 F. Supp. 2d 1070 (C.D. Cal. 1999).

Porsche Cars North America v. Porsch.com, 51 U.S.P.Q.2d 1461 (E.D. Va. 1999).

Quilt SRO, Regional Court (Pilzen, Czechoslovakia, 2001).

Sallen v. Corinthians Licenciamentos, Ltd., 2001 U.S. App. LEXIS 25965 (1st Cir. 2001).

Sankyo Co. Ltd. v. Zhu Jia Jun, WIPO Arbitration Case No. D2000-1791 (2000).

Smart Design, LLC v. Carolyn Hughes, Case No. D2000-0993 (2000).

Toys R Us v. Mohamed Akkaoui, 1996 U.S. Dist. LEXIS 17090 (N.D. Cal. 1996).

U.S. Anticybersquatting Protection Act (Public Law No. 106-113) (1999), 15 U.S. C. Sect. 1125(d)(1).

WIPO Arbitration Forum, <www.arbiter.wipo.int/domains>.

Zurakov v. Register.com, N.Y. Sup. Ct., N.Y. Cty., Case No. 600703/01 (July 25, 2001).

5

Protecting the look and feel of online content

In this chapter, we will discuss the challenges associated with asserting ownership rights over the appearance of online content, including the on-screen appearance of computer software and Web pages. As noted in other chapters of this book, there are several different strategies applicable to the protection of computer code and information made available online. To date, however, protection of the on-screen appearance, the "look and feel" of online content, has been more difficult to accomplish. This chapter addresses those traditional difficulties and offers strategies that can be used to provide at least some level of control over the appearance of online content. The topic of effectively managing the visual appearance of a software or online user interface or transaction system is increasingly important as the recognized commercial value of that electronic material grows. Protecting electronic look and feel is thus a subject that is highly dynamic and is likely to remain so in the future.

Principles for protection of the appearance of tangible products are well established, and that protection has been provided under concepts of trademark, patent, and copyright law. For example, the legal theory of trade dress has long provided effective legal rights for the developers of highly recognizable product packaging (e.g., Coca-Cola's distinctive bottle shape). That protection for the appearance of packaging has been broad enough to protect both the precise form of the package and the general look and feel of the package (e.g., trade dress principles permit Coca-Cola to protect its novel bottle shape no matter what color is applied to the bottle). Although the legal theories appropriate for protection of ownership rights for look and feel are fairly well developed as to tangible products, they remain far less clear with regard to digital products.

5.1 Limits of copyright protection

Copyright law provides only limited protection for the look and feel of online content. As noted previously, copyright law provides ownership rights for the authors of computer programs. Thus, to the extent that computer programs create a specific user interface or other form of on-screen content, ownership rights for those programs are established and enforced by copyright law. Those rights are limited, however, since they apply primarily to the code and not as comprehensively to the on-screen appearance of the content. This distinction is important, since it means that another party is may be free to simulate the on-screen appearance, provided that it does not infringe on the copyright for the original computer code to recreate that appearance. Copyright law protection for the "literal" elements of computer software (i.e., the source code and the object code) is strong. Copyright law protection for the "nonliteral" elements of computer programs (e.g., the visual representation, the user interface) is inconsistent.

Copyright law thus provides only limited protection for developers of innovative on-screen interfaces (e.g., graphical user interfaces) and content appearance. If a developer of online content independently creates material that is similar to the look of the content created by another party, both parties have the right, under copyright law, to assert ownership over their respective works. Also, if a developer creates a similar look for online material through use of a lawful, properly performed, process of reverse-engineering, there is no copyright law violation. Thus, even if a particular form of visual appearance is within the scope of copyright law protection, that protection does not provide enforceable rights in the event of truly independent creation of a similar work.

In some jurisdictions, copyright law applied to computer programs explicitly excludes user interfaces from copyright protection. For instance, Japan does not apply copyright law protection to user interfaces. This exclusion is designed, in part, to prevent ownership conflicts from impeding interoperability and widespread adoption of software and online content. Such an approach does, however, make it more difficult for content developers to establish a unique appearance for their digital material. Copyright law in the European Community applies a similar approach to user interfaces and software interoperability. Although copyright law in Europe expressly protects computer programs, the law also permits access to the underlying code for purposes of ensuring user access and effective interoperability among diverse software systems. In the United States, however, applicability of copyright law to user interfaces is less clear. Copyright law in the United States permits developers to assert rights in user interfaces provided that the developers can demonstrate that the interface is original and reflects substantial creativity. Interfaces based on public domain or other preexisting software, and those that do not demonstrate significant creativity and originality, are not suitable for copyright protection.

Copyright law can provide protection for discrete elements of online content. For example, text, photos, audio recordings, and graphic elements are all protected by copyright law. Copyright claims associated with material made available online are common. This protection has limits, however. It prevents another party from duplicating the elements of an online presentation, but it does not prevent independent creation of a similar presentation style. Copyright law also prevents the embedding of copyrighted material into the Web content of another party, unless prior consent has been obtained from the owner of the embedded work. Copyright law has also been used in efforts to block hypertext linking between Web pages. Courts have been reluctant to find hypertext links to be copyright law violations, except in those instances when deep links are involved (i.e., links to pages other than the home page), and when the links resulted in quantifiable economic harm (e.g., reduced advertising revenues) to the site objecting to the link.

Look and feel and other nonliteral elements of computer software and online content are protected by copyright law only to the extent that they display creativity and originality. Nonliteral elements that can be traced back to public domain or widely applied content are not generally protected. Similarly, nonliteral elements that were motivated primarily by external factors, such as operating efficiency, instead of creativity, are generally granted less protection under copyright law than those works inspired more clearly by a desire for creative expression. In addition, facts and information that are collected, but are not works of creativity, are not subject to copyright law protection. Thus, the telephone number listings contained in telephone directories, which were the subject of the *Feist Publications, Inc. v. Rural Telephone Service* case, were not granted copyright law protection, since the court determined that telephone numbers were facts, not creative works of authorship.

5.2 Patents and digital look and feel

In a previous chapter, we discussed the fact that patent law provides a means through which developers of computer programs can assert control over use of their software. Computer programs can be protected under patent law; however, the scope of that protection varies from country to country. Software patents are widely applied in the United States but are far less comprehensively accepted in Europe, Japan, and other parts of the world. Patent law has not, to date, been widely applied to management of the look and feel of online content. Several aspects of patent law may at some point, however, come into play with regard to the appearance of digital content. Those aspects are the U.S. patent law "doctrine of equivalents," the scope of business method patents, and the application of design patents.

At the most basic level, patents prevent unauthorized parties from developing, manufacturing, distributing, or using devices that fall within the scope of the "claims" for the protected invention that are established in the patent. The claims include descriptions of the invention itself and of the functions the invention performs. There is patent infringement if the unauthorized party creates or uses a device that falls within the scope of the claims for the patented invention. In addition, U.S. courts applying a principle known as the "doctrine of equivalents," are also willing to find patent infringement, at times, even if there has been no direct violation of a specific patent claim. The basic form of patent infringement involves "literal" infringement. Literal infringement occurs when every element that is described in a patent claim is found in a device created by another party. The doctrine of equivalents asserts that a court may find patent infringement when there is no literal infringement. The doctrine states that, even if the infringing device does not contain all of the elements defined in a patent claim, a court can find infringement. This nonliteral infringement exists if the differences between the device and the patent claims are insubstantial and if the unauthorized device accomplishes substantially the same function as the patented device, in a manner substantially similar to that used by the patented device.

The doctrine of equivalents may extend patent protection to a degree sufficient to provide some future protection for the look and feel of digital content. If a user interface or online transaction process is included within the scope of the claims for a software patent, violation of those claims by another party would constitute patent infringement. In addition, under the doctrine of equivalents, even if the actual code creating the interface or transaction system did not directly infringe on the patent of another party, there could nonetheless be patent infringement if a court concluded that computer programs similar to the protected programs were used to accomplish substantially the same function as the protected code in substantially the same way. While conducting an analysis of the similarity between the programs, as to form and function, courts will consider nonliteral aspects of programs, including their look and feel. In this way, similarities as to look and feel, and other nonliteral elements of the computer programs, can contribute to a court determination that there has been patent infringement.

A recent U.S. federal court case restricted the application of the doctrine of equivalents somewhat (*Festo Corporation v. Shoketsu Kinzoku Koygo Kabushiki Ltd.*). The court in the *Festo* case concluded that the doctrine of equivalents applies only to the claims made in the original patent application. At issue in the *Festo* case was a patent for a rodless cylinder tube, which contained a magnetic piston that could function as a magnetic clutch. Under *Festo*, claims that are modified or added through amendments to a patent application are not subject to the doctrine of equivalents; thus, the patent protection provided to supplemental or amended patent claims is not as broad as that granted to the original claims in patent. In spite of the *Festo* case, however, the doctrine of equivalents remains a valid and important aspect of U.S. patent law.

Business method patents, particularly when applied in the context of electronic commerce systems and processes, provide an illustration of the ways in which patents can provide protection that extends beyond computer programs and includes user interfaces. Federal courts in the United States and the U.S. Patent and Trademark Office now recognize that certain forms of methods of conducting business activities and functions can be protected under patents. The key U.S. court case establishing this principle was *State Street Bank and Trust Company v. Signature Financial Group, Inc.* (1998). This patent protection applies to the computer programs used to automate business methods and to the methods themselves. Business method patents assert ownership rights over business activities or functions. For example, in the *State Street* case, the patent covered methods of allocating funds among different financial investment instruments, with a goal of maximizing the financial gain from the fund allocation. Many different business methods have now been patented, including a variety of e-commerce business systems (e.g., Amazon.com's "one-click" online payment system, Cybergold's "attention brokerage" online marketing system, and Priceline.com's buyer-driven online auction system). These e-commerce business method patents remain controversial, and there have been court actions challenging the validity of some of those patents (e.g., *Amazon.com v. Barnesandnoble.com*).

The e-commerce business method patents currently enable the developer of an electronic business method system to assert control over both the underlying computer code that enables the system and over the user interface with that system. Commonly, business methods patents are structured to include both the code and the functionality driven by the code within the scope of the patent claims. For example, the Amazon.com one-click payment patent encompasses both the software that drives that payment system and the user interface with the system. The computer program and the transaction process automated by the program are both incorporated into the scope of the e-commerce business method patents. Thus, the U.S. Patent and Trademark Office determined that the software and the process associated with a single-step payment processing system was patentable.

Based on the grant of that patent, Amazon.com sued Barnesandnoble.com for infringement, arguing that the payment system used by Barnesandnoble.com on its Web site infringed on the Amazon.com patent. In that case, the trial court granted a temporary injunction for Amazon.com, barring Barnesandnoble.com from using its one-step system, pending the resolution of the infringement lawsuit. Barnesandnoble.com appealed the grant of the temporary injunction, and the appeals court overturned the injunction and ordered the trial court to consider the validity of the Amazon.com patent. The parties agreed to a settlement before the trial court could complete its examination of the Amazon.com patent. The question of the scope of business method patents for e-commerce processing systems remains uncertain. Clearly the PTO and the courts recognize that such patents can be valid; however, the exact scope of business method patents in the electronic commerce environment remains uncertain, since no court has yet had the opportunity to define them fully.

This provides a clear example of the way in which a business method patent can control both the software aspect of the process in question and the user interaction system associated with the software. In effect, these business method patents apply both to the code that drives the system and the look and feel of the system, from the perspective of the user. Although there remains substantial uncertainty as to how far the courts will be willing to go to enforce these patents and whether countries other than the United States will ultimately embrace these patents, the value of business method patents to e-commerce system developers is quite clear. Business method patents in the e-commerce environment illustrate the broad potential scope of coverage that patents provide, and they demonstrate how patent protection can sometimes provide far more comprehensive coverage for look and feel than can traditional copyright protection.

The debate associated with business method patents also illustrates the extent to which patent protection can limit application of new technologies. Opponents of business method patents fear that widespread use of those patents can impede expansion of innovative e-commerce systems. Certain systems supporting online commercial transactions and e-commerce functions are widely used and are essential to many different enterprises conducting online business functions. Some industry observers express concern that aggressive business method patent enforcement coupled with restrictive licensing strategies will make it difficult for diverse parties to make use of these key technologies and will thus impede e-commerce expansion. Proponents of business method patents contend that those patents provide important economic incentives for continuing innovation and that competitive concerns can be readily handled through use of reasonable licensing systems (e.g., compulsory licensing requirements). This debate is far from resolved and will likely continue well into the future.

Another aspect of patent law that has a direct bearing on management of look and feel of content is the scope of rights associated with "design patents." The most common form of patent is the "utility" patent. Utility patents protect inventions that have useful functions. There is, however, another form of patent known as a design patent. Design patents protect the appearance of inventions. Applications for design patents are not required to demonstrate a useful purpose for the invention; instead, they must show that the proposed design has a primarily ornamental function and is novel. Infringement of a design patent occurs when an unauthorized party creates or makes use of a design that is identical or substantially equivalent, in an ornamental sense, to the patented design. Design patents protect only the appearance of the device, not the device itself. For example, a design patent could be obtained to protect a specific design for a chair, but that patent would not protect the chair as a device. Design patents are commonly used to supplement utility patents. In the chair example, for instance, the inventor of a chair that has both novel functions and design can seek a utility patent to protect the novel functions of the chair and a design patent to protect the innovative appearance of the chair.

Although design patents have been recognized in the United States for many years, design patents for computer software have not been as widely applied as one may have expected. This delay in applying design patents to visual aspects of computer programs was caused primarily by the traditional patent law view that design patents applied only to manufactured items. Under this old doctrine, design patents were commonly awarded for the design of computer equipment; however, they were not granted for on-screen content (e.g., Ex parte Donaldson [1993], Ex parte Strijland [1993]. The old doctrine was eventually modified in several patent cases in the United States, including Ex parte Donoghue (1993), where the U.S. Patent and Trademark Office Board of Appeals suggested that icons presented on computer screens could be subject to design patent protection. That position was ultimately formally clarified in 1996 by the U.S. Patent and Trademark Office in its *Guidelines for Examination of Design Patent Applications for Computer-Generated Icons.*

Design patents are also becoming more widely recognized for electronic content outside of the United States. For example, Canada liberalized its use of design patents to protect on-screen icons and other forms of electronic visual content. The Canadian Industrial Design Office expressly recognized the validity of design patents for on-screen icons in 1997. Design patent registration is now commonly used for on-screen icons in Canada.

Although the primary focus of design patents, to date, has been protection of distinctive on-screen icons, there may be greater use of design patents in the future, as part of a more aggressive effort to protect digital look and feel. For example, a software developer could apply for a utility patent for the novel computer code he or she developed and could apply for a design patent to protect the overall on-screen representation of the program, not simply the form of the icons. Design patents can be used to supplement utility patents for computer programs, and that combination of patents can provide additional protection for the look and feel of digital content. As noted in our previous discussion of software patents, however, recognize that the process of obtaining patent protection for software is generally slower, more cumbersome, and more expensive than other intellectual property protection strategies; thus, patents may not be suitable for all content or all content developers.

5.3 Trademark law protection

Trademark law offers some potential protection for the look and feel of online content. As noted in the previous chapter, trademark law provides protection for commercial identifiers that are used in the marketplace to identify a specific company or product to consumers. Trademark law protects the investment that businesses make in creating these commercial identifiers. To date, trademark law has been used primarily in the context of domain names and keywords in the online environment. It is likely that trademark law theories discussed previously in this book (both misappropriation and dilution) will be increasingly popular tools to assert control over the general appearance of online content in the future.

Trademark law has been most widely applied to protect specific elements of online material. Words, logos, and other specific trademarks are commonly carried over from the brick and mortar world to the digital environment. All of those marks are subject to the same degree of trademark law protection in the digital environment that they receive in the physical world. Trademark law enables the developers of distinctive commercial identifiers to prevent other parties from using marks that can lead to consumer confusion. Trademark law rights have been widely enforced in the electronic marketplace.

In addition to protection of specific trademarks, established in the digital environment, there are also examples of trademark controversies associated with the development of electronic versions of traditional products. For example, Minnesota Mining & Manufacturing Company (3M) was at least temporarily in conflict with Microsoft Corporation over electronic versions of the "Post-it" note product developed by 3M. Microsoft developed a feature as part of its Windows product that enabled users of the software to create and use digital equivalents of those notes. This dispute presents an example of possible future controversies as electronic versions of established goods develop, and the owners of the trademarks for the brick and mortar versions of the products attempt to enforce their rights against the developers of the electronic versions.

The overall look and feel of digital material has not yet been widely recognized as trademarked material. To the extent, however, that the appearance of on-screen content becomes identified by consumers, with a particular organization, one can argue that the look and feel of that material is a trademark. To qualify for this protection, however, the look of the content must be widely recognized by the public and it must be presented in the same way at all times. Variation in a mark undermines the ability of the user of the mark to assert trademark rights over the mark. Thus, if one intends to assert trademark rights over the on-screen presentation of material, that presentation must always be made in the same format. Variation of the on-screen look could eliminate the trademark right. This consistency of presentation requirement poses a challenge for those who would attempt to apply trademark law theories to restrict creation of online content that has a similar appearance, since online content is frequently modified.

Another aspect of trademark law that can come into play in this setting is the concept of trade dress. Trade dress rights are established by Section 43(a) of the U.S. Lanham Act. In most instances, trade dress protection is applied to the design of packaging associated with a company and its products. Distinct designs for containers and other forms of packaging or wrappers are protected under the trade dress principles of trademark law. Trade dress protection can extend beyond pure packaging. For example, products that have a unique and distinctive design (e.g., a specific design for a piece of furniture) can be protected under trade dress arguments. Extending this trade dress application to the digital world, it may be possible to apply a trade dress theory to assert control over some forms of online content appearance or presentation style.

The trade dress argument in the context of look and feel for digital content can be built around an analogy between on-screen, visual representation and tangible product packaging and design. This analysis involves a claim that the look and feel of digital material performs a function similar to that of packaging for tangible goods and similar to the design of a tangible product. The trade dress argument is increasingly common as competition for distinctive look and feel increases. To raise this claim successfully, the developer of software or online content must prove three facts. The developer must show that the material claimed as trade dress has developed a secondary meaning—that is, the material is now identified by consumers with the developer. The trade dress claim also requires proof that the allegedly infringing use of the material by the unauthorized party is likely to cause consumer confusion. Finally, the developer must demonstrate that the material that has been used has a purpose that is ornamental, not functional. Trade dress rights are intended to protect the ornamental value of the material, not its functional value. Thus, for example, if the on-screen representation in question was motivated for functional purposes (e.g., cost reductions, ease of customer use) instead of ornamental purposes, the material may not qualify as trade dress.

5.4 Rights under the Database Directive and property law theories

The Database Directive of the European Community established a framework for assertion of ownership rights over digital material that does not qualify for traditional copyright law protection. As discussed in greater detail in Chapter 8, the European database rights are primarily intended to enable the developers of data collections to protect their investment of resources applied to the creation, maintenance, and updating of those collections. Although the extent of coverage provided by the directive is not yet fully defined, it has been interpreted to permit assertion of ownership rights as to online content, including Web pages. This interpretation opens the door for protection of online content appearance under the scope of the directive. For example, a German court noted that a collection of Web pages can be protected as a database under the Database Directive, in the *baumarkt.de* case. In that case, the court found that there had not been a substantial enough investment of resources applied to the creation of the Web pages to merit an award under the directive; however, the court accepted the concept that unauthorized application of frames to online content could be a violation of the ownership rights established by the directive. Another German court, in the *Medizinisches Lexicon* case, also accepted the premise that framing, as well as linking, without the authorization of the developer of the Web content, can be a violation of the database ownership rights established by the directive.

These cases suggest that database ownership rights established in the European Community can, under some circumstances, be used by Web content developers to protect the visual representation of their content. It is important to recognize that the rights established under the directive are essentially property rights, not traditional intellectual property rights. To the extent that online content falls within the scope of the Database Directive, the directive provides an additional tool for use by online content developers to manage their online material. If the look and feel of electronic content that falls within the scope of the directive is duplicated, without authorization, it seems that the developer of the original content could raise a legal claim under the Database Directive. Such a claim would be in addition to any traditional intellectual property law claims that might arise.

5.5 Trespass to property

An increasingly common legal claim applied to control access to, or use of, digital content is the claim of "trespass" to property. Well-established legal principles in many jurisdictions enable the owner of land or other forms of tangible property to limit the ability of other parties to use that property. These trespass theories are discussed in greater detail in Chapter 8, but they are also relevant in this discussion of management of the appearance and visual representation of electronic content. Trespass arguments have been raised as barriers against parties attempting to make use of a wide range of digital content. For example, ISPs have successfully applied trespass arguments against parties who direct mass mailings of unsolicited commercial e-mail messages to the ISP network (*CompuServe, Inc. v. Cyber Promotions, Inc.* [1997]). Trespass arguments have also been raised against parties using search engines to access commercial information stored on another party's server (*eBay v. Bidder's Edge* [2000], *Register.com v. Verio* [2000]).

In these cases, the trespass argument is raised on two levels. On one level, the property involved is deemed to be the computers that are accessed by the unauthorized party. The spammer directs the messages to the ISP's network, thus "using" the ISP's computers (i.e., property). The search engines access the servers where the target information is stored; thus, they too access the computer equipment that is the property of the other party. Under this line of reasoning, the computer equipment is property, and it has been used in a manner inconsistent with the limited usage authorization granted by the owner. Additionally, in some of these cases, another form of property is accessed. That second form of property is the data files themselves. Property trespass arguments can thus be based either on a claim that a party has made use of a computer without adequate permission or that the party accessed a data file without suffi-cient authorization. In both instances, property of one party has been used in ways that exceed the permission granted by the property owner.

Under certain circumstances these property trespass claims could be applicable in the context of duplicated look and feel, as well. If it can be proven that a party accessed a computer or data files owned by another party, without permission, and if the trespassing party created a look and feel for electronic content similar to that of the original party's material, it seems that trespass claims could be raised by the developer of the original content. This theory has not yet been widely applied to disputes regarding duplication of visual appearance of content; however, it seems that the theory is as valid in this context as it is in those in which it has already been raised. Note that this claim is independent of any intellectual prop-erty law rights. To raise this property argument, the developer of the orig-inal content must prove ownership and unauthorized use. That developer would not be required to prove any intellectual property rights, and the defendant in such an action would not be able to employ any of the tradi-tional intellectual property law defenses, such as fair use.

5.6 Rights under competition law

Competition law prohibits unreasonable actions that result in harm to commercial competition. A claim of unfair competition can be sustained if the conduct involved is unreasonable and results in harm to competi-tion in a commercial market. It is possible to use competition law as a means to protect commercially valuable material, even if that material does not fall within the scope of traditional intellectual property or gen-eral property law principles. Competition law rights are not designed to protect property but are instead directed toward preserving a commercial environment that supports effective business competition.

Duplication of visual representation format can constitute unfair competition. In order to sustain such a claim, the developer of the original material must persuade a court that the creation of a similar look and feel by another party has unfairly harmed commercial competition. Part of that showing involves documentation of economic harm caused by the duplication; however, a showing of such economic harm alone is not suffi-cient to support the claim. In addition, the plaintiff must demonstrate that the conduct of the defendant was unreasonable and that the conduct has reduced the future prospects for competition.

5.7 Moral rights

Certain countries, particularly those in Europe, have a long tradition of legal rights for artists to protect the integrity of their creative work. Described as moral rights ("droits morale"), they consist of rights that are only enforceable by developers of artistic material. The rights are limited to the ability to prevent alteration or use of the artistic material in ways that adversely affect the artistic integrity of the work. Moral rights focus on protection of an original work from harmful alteration, they do not focus on creation of derivative works. Traditional copyright law provides the vehicle to block unauthorized derivative works. The principle of moral rights was accepted as part of the Berne Convention for the Protection of Literary and Artistic Works (Article 6). To date, moral rights have not been aggressively asserted with respect to digital media; however, as online content becomes more widely embraced, it is possible that moral rights arguments may be applied by developers of online creative material.

If applied in the context of digital content look and feel, moral rights could be used by the developer of that content to prevent others from altering the visual appearance of the content. Moral rights would not, however, be effective to prevent another party from creating digital content that had a similar appearance, unless the creator of the original work could demonstrate that the creation of the duplicate work undermined the artistic integrity of the original work. Moral rights do not automatically prevent creation of similar works; instead, they provide a legal support to prevent actions that harm the creative integrity of the work in question. Moral rights can only be asserted by the individual who actually created the original work, since moral rights are not transferable to another party (e.g., an employer). In addition, the only remedy that is generally provided by moral rights is a court order requiring the defendant to stop the action that harms the integrity of the work. Moral rights do not provide for monetary damages as compensation or as a penalty.

To enforce moral rights as to the look and feel of electronic content, the creator of the original content must persuade a court that the content was artistic in nature. The creator must also prove that the conduct of the defendant harmed the artistic integrity of the original work. Moral rights claims have not yet been widely applied as to electronic content. The potential applicability of those claims as to digital material is currently difficult to assess; however, it appears that there could be factual circumstances that could sustain moral rights claims associated with the look and feel of electronic content, although such claims are not likely to be widespread.

5.8 Managing look and feel

Traditional legal theories provided limited protection for the visual representation of online content. Copyright law has been the primary legal tool applied to protection of electronic content; however, it has had limited and inconsistent applicability for on-screen content appearance. Traditional copyright law has been more useful protecting the underlying computer code and the specific elements of digital media content than the overall look and feel of the electronic presentation. As the commercial value of the visual representation of electronic material has increased, developers of that material now look for additional legal principles that will enable them to assert greater control over access to, and use of, their creative content.

In this environment of growing concern as to the need to protect unique content presentation formats from duplication by competitors, certain aspects of traditional intellectual property law will likely be increasingly popular in the future for use in the process of managing the visual appearance of digital media content. For example, patents are increasingly common for computer programs, and it is likely that both utility and design patents will be more actively applied to manage on-screen content appearance. Increased use of business method patents also provides an additional vehicle for assertion of control over certain commercial interfaces and processes in the electronic marketplace. Trademark and trade dress arguments may also become more visible as online content continues to develop expanded commercial importance.

New legal theories and unconventional application of some traditional legal principles, in addition to intellectual property laws, appear to provide promising mechanisms to manage on-screen content appearance for the future. Database ownership rights established in Europe are, at least in part, now applied to online content and appear to provide a legal basis for control over duplication of on-screen look and feel. Laws designed to protect traditional property rights and to promote fair commercial competition also provide tools applicable for developers of innovative forms of visual presentation for digital content who seek to protect their rights. Moral rights granted to artists may also have a limited role in future management of digital look and feel.

In spite of this diverse range of legal principles supporting assertion of control over the appearance of electronic content, developers of that content should not assume that they will be able to prevent creation of visual representations similar to their original works for an indefinite period of time. As content owners map their strategies for the development and distribution of their content, they should not base those strategies on an assumption that their content will always have a unique appearance. It is a mistake for creators of content to believe that they will be able to make use of legal rights to preserve the unique nature of the look and feel of their digital content for an extended period of time.

Although they cannot protect their creative advantage indefinitely, creators of content should nevertheless act prudently to protect the look and feel of their material. They can best accomplish this task by applying multiple legal strategies to establish and enforce controls on duplication or other use of the visual representation of their material. For instance, use of copyright, trade dress, and design patent rights for the same on-screen material can provide a broader scope of protection for that material than would reliance on a single legal claim. Yet even as those developers judiciously preserve and enforce the diverse legal tools available to them for management of the look and feel of their material, those developers should recognize that the law can never provide absolute protection for any competitive advantage they may derive from innovation.

Content creators should also make sure that they respect the integrity of the appearance of material developed by others. In the increasingly commercial setting in which digital content is developed and used, most parties are both creators and consumers of intellectual property. When in the role of content developer, all parties seek to maximize the benefit they derive from their creative work. In the role of consumer of creative content, however, parties should recognize that they face potential legal liability for misuse of the property of others. The increasingly diverse theories likely to be applied to management of look and feel in the future will make it more challenging for consumers of electronic content. Accordingly, all users of that content should respect the rights of others and reduce their liability exposure by understanding the different rights and obligations associated with content presentation and by developing policies and practices that foster compliance with all applicable requirements.

Selected bibliography

Amazon.com v. Barnesandnoble.com, Case No. 00-1109 (Fed. Cir. 2001).

Amazon.com One-Click E-Payment System, U.S. Patent Number 5,960,411.

Apple Computer Corporation v. Microsoft Corporation, 35 F.3d 1435 (9th Cir. 1994).

baumarkt.de, Oberlandesgericht (Court of Appeals) Dusseldorf, 1999.

Berne Convention for the Protection of Literary and Artistic Works, at <www.law.cornell.edu/treaties/berne/overview.html>.

Canadian Intellectual Property Office and Canadian Industrial Design Office, "Guidelines for Registration of Electronic Icons," (1997), at <http://cipo.gc.ca>.

CompuServe, Inc. v. Cyber Promotions, Inc., 962 F.Supp. 1015 (S.D. Ohio 1997).

Computer Associates International, Inc. v. Altai, Inc., 23 U.S.P.Q 2d 1241 (2d Cir. 1992).

Cybergold Attention Brokerage Online Marketing System, U.S. Patent Number 5,855,008.

eBay v. Bidder's Edge, 100 F. Supp.2d 1058 (N.D. Cal. 2000).

Ex parte Donaldson, 26 U.SP.Q. 2d 1250 (1993).

Ex parte Donoghue, 26 U.S.P.Q. 2d 1266 (1993).

Ex parte Strijland, 26 U.S.P.Q. 2d 1259 (1993).

Feist Publications, Inc. v. Rural Telephone Service, 113 L. Ed. 2d 358 (1991).

Festo Corporation v. Shoketsu Kinzoku Koygo Kabushiki Ltd. (Fed. Cir. 2000).

HOB Entertainment, Inc., et al. v. Streambox, Inc., Case No. 00-10017 (C.D. Cal. filed Sept. 13, 2000).

Lanham Act of 1946, Section 43(a).

Lotus Corporation v. Borland Corporation, 49 F.3d 807 (1st Cir. 1995).

Medizinisches Lexicon, Landgericht Hamburg, 2000.

MercExchange Computerized Market-Maker Online Auction System, U.S. Patent Number 5,845,265.

Priceline.com Buyer-Driven Online Auction System, U.S. Patent Number 5,794,207.

Register.com v. Verio, 126 F.Supp. 2d 238 (N.D. Cal. 2000).

Schoolhouse, Inc. v. Jeff Anderson, et al., Case No. 00-3939 (8th Cir. 2001).

State Street Bank & Trust Co. v. Signature Financial Group, Inc., Case No. 96–1327 (Fed. Cir. 1998).

Ticketmaster Corp. v. Tickets.com (C.D. Cal. March 27, 2000).

U.S. Patent and Trademark Office, "Guidelines for examination of Design Patent Applications for Computer-Generated Icons," Docket No. 950921236-6049-03, RIN 0 651-XX04 (March 21, 1996), at <www.uspto.gov/web/offices/com/sol/notices/icon.html>.

Whelan Associates, Inc. v. Jaslow Dental Laboratory, Inc., 230 U.S.P.Q. 481 (3d Cir. 1986).

Xerox Computer Display, U.S. Patent Number 296,339 (1988).

6

Managing digital media

As more diverse media content is widely distributed online in digital form, management of use of that material becomes a greater challenge. Of particular importance is the need to reconcile digital media rights with rights of access and use already established in the offline marketplace. Print publishers, the music industry, and the video industry (including both television and motion picture producers) all face complicated issues associated with the management of digital media rights. In this chapter, we will examine some of those key issues and discuss possible solutions.

6.1 Print publishers

Book and magazine publishers were among the first media content industries to face the challenges associated with the rise of digital media. Conflicts arose as to the rights of traditional print publishers with respect to digital versions of their content and as to the rights of those publishers relative to the rights of purely electronic publishers. In general, traditional print publishers have tried to extend their publication rights into electronic formats, while electronic publishers attempted to preempt those electronic publication rights. At the same time, authors have fought to

AuThoR PUBLISHER DISTRIBUTOR RETAILER CONSUMER

preserve their ability to control the distribution of their works and to obtain fair compensation, in both traditional and new media forms.

In the world of print publication copyrights, there are several key players. The authors create the works. Publishers (both book and periodical publishers) contract with authors, and the publishers handle the production and distribution of the works. Distributors make publications available to retail outlets, which, in turn, sell the publications to the public. Copyright in a written work first rests with the author. Many publishers, however, insist that the copyright be conveyed by the author to the publisher. Traditionally, either the author or the publisher controls the copyright for the work, and none of the other participants in the distribution chain possesses any copyright claims as to the publication. Rights associated with print publications are becoming more complex, as an increasing amount of published content is available in both traditional print and electronic formats.

The court case of the *New York Times v. Tasini* illustrates the ongoing efforts to clarify the rights of authors and publishers as to electronic versions of published works. In that case, authors challenged the right of print publishers to include works created by the authors in electronic databases independent of the original collections in which those works were published. The *New York Times* and the other publishers involved in the case had clearly purchased the right to publish the works of the authors in print collections. However, in addition to the print publications, the publishers included the works of the authors in electronic databases, without providing any additional compensation to the authors. In those electronic databases, the works of the authors were accessible individually (i.e., separate from the original collections of printed works in which they had been originally published).

The authors claimed that because their works were now accessible to readers on a stand-alone basis, separate from the original collections, the publishers should be required to obtain additional permission from the authors for those additional publication rights. The publishers argued that no additional permission should be required, since the publication of the works in the electronic format should be included in the original grant of print publication rights. The case ultimately went to the U.S. Supreme Court, and that court ruled in favor of the authors. The Supreme Court concluded that when an author's work is accessible in an electronic database in isolation from the original collection in which it was published, that publication is no longer part of the publication of the original collection; thus, the author retains the right to publish the work in that isolated, electronic format, and the publisher must obtain additional permission and provide additional compensation, as appropriate.

Another court case in the United States addressed the issue of the extent to which print publication rights extend to creation of electronic versions of printed material. The case was *Random House, Inc. v. Rosetta Books*. In that case, the electronic book publisher, Rosetta Books, asked the court to clarify whether contracts that book publishers had entered into with authors granted publication rights for electronic versions of books. Rosetta Books argued that the older standard publication contracts that Random House and other book publishers had entered into with their authors applied only to publication of books in hard copy printed form, not to publications in electronic form. The publishers argued that the contract definition of books was broad enough to include electronic books, and thus the contracts gave the traditional publishers exclusive rights to publish the work of their authors in both print and electronic form. The court ruled in favor of Rosetta Books, interpreting those older contracts to apply only to print versions of the books, not to electronic versions. The publishers have appealed the trial court's ruling in the case.

The *Tasini* and *Rosetta Books* cases illustrate the need to recognize and to clarify the grant of digital publishing rights when handling the distribution of traditional content in electronic formats. Publication rights granted in one medium will not automatically provide legal rights adequate to support republication of the material in electronic form. Digital publication rights should be specifically addressed in any license or other transfer of publication or distribution rights. In addition, to the extent that digital publication rights are separate from traditional print publication rights, transfer of those rights will generally involve an additional increment of compensation through royalties or other license fee structure. Also, these cases underscore the fact that electronic formats for published material commonly facilitate nontraditional means of access and use for the material (e.g., the ability to access individual works from collections of material, which was at the center of the *Tasini* Case). Those nontraditional aspects of electronic formats require special attention as the relative rights of print and electronic publishers, and authors, are established.

6.2 Music rights

The music industry has engaged in the most visible struggles related to assertion of content rights in the online setting. Music rights are complicated, since there are several different parties involved in the creation and distribution of music. There are several key participants in the music distribution chain. Composers of music are the creators of the content, and all copyright rights in musical compositions initially rest in their hands. Music publishers contract with composers, and those publishers create and distribute the print versions (scores) for the musical compositions. Composers contract with performance rights organizations (e.g., the American Society of Composers, Authors and Publishers [ASCAP] and Broadcast Music, Inc. [BMI]), and those organizations represent the composers to make sure that all who perform the musical compositions created by the composers pay an appropriate royalty to the composer. Artists who perform works in live concerts must obtain performance rights from the composer or from the appropriate performance rights organization. Record companies that create recordings of music must obtain recording (or mechanical) rights from the composer. Broadcasters (e.g., radio stations, television stations) and other parties who play recorded music for the public (e.g., motion picture producers, restaurants, and other public venues) must obtain permission for those performances from the composer or from the appropriate performance rights organization. Copyright law in the United States does not require prior permission for public performance of recorded works, based on the rationale that such public performance of recordings (e.g., broadcast of a recording on the radio) serves to promote the recording and thus increase the future sales of that recording. The Digital Millennium Copyright Act, however, permits artists and record companies to collect royalties for recorded music that is performed on the Internet. That provision is controversial and has led to substantial debate over the size of the royalty to be assessed. This debate is discussed more fully later in this chapter.

During the early stages of the digital rights debate in the music industry, one of the basic questions to be resolved was the issue of how to characterize accurately the form of online music distribution. The music industry traditionally divided music rights into several basic categories: publishing rights, recording or mechanical rights, and performance rights. Different licensing systems developed for each of those basic music rights categories. In the early days of online music distribution, there were debates as to which of the categories of music rights were affected by music made available on the Internet. Over time, it has become clear that the answer to the question regarding which of the traditional music rights apply to online music distribution is that all of those rights apply. Music publishing rights apply when use of the printed scores for music are involved. Music recording rights apply when sound recordings of music are made available online. Performance rights for music apply when the music is actually "played" online.

The online music rights debate is made more complicated by the fact that so many different players in the traditional music development and distribution chain are now actively involved with music in digital format, and nontraditional participants (e.g., "Webcasters") have also entered the field. For example, music publishers now routinely make their print scores available online. Music recording companies offer their versions of online distribution of recordings by their artists. Some musicians distribute their music directly to consumers online. Music broadcasters (e.g., radio stations) now commonly provide access to their broadcasts via the Internet. This constellation of traditional music distributors who are now operating online is supplemented by pure online music distributors who now participate in the marketplace. Examples of these exclusively online music distributors include Webcasters (i.e., the online equivalent of radio broadcasters) and online music exchange systems (e.g., Napster, which will be discussed in greater detail in Chapter 7). The primary distinction between the Webcasters and the online music exchange systems is that the Webcasters parallel traditional broadcasters, while the online music exchange systems parallel true music distributors. Webcasters make use of streaming media formats that do not transfer possession of control of the music file to the listener. Online music exchange systems transfer possession and control of the music file to the recipient; thus, the recipient is then in possession of a digital file that can be replayed, duplicated, and redistributed. Leading Webcasters include Clear Channel and BMG Music Service. Prominent music exchange systems include Napster, MusicNet, and Pressplay.

There are several examples of the conflicts caused in the music industry by the expansion of online music distribution. For instance, the organizations that control music publishing rights are pushing the music recording companies to pay additional royalties for music distributed online. Music publishers distribute musical scores, not sound recordings; thus, music publishers do not record or perform musical works. In the United States, the leading music publishing rights organization is the

Harry Fox Agency, a unit of the National Music Publishers Association, which enters into the licensing agreements, on behalf of music composers, that authorize use of musical scores. The Harry Fox Agency manages access to the published scores of musical compositions. The agency is presently negotiating with music recording companies to obtain additional royalties for recorded music that is distributed on the Internet. The music publishers argue that those royalties should be in addition to the already established royalties associated with recording of the published music and with broadcast distribution of the music. The record companies disagree, contending that the online distribution system is too immature yet to enable the parties to identify an appropriate royalty fee. The dispute has been raised both in court (*The Rodgers & Hammerstein Organization v. UMG Recordings, Inc.*, a lawsuit brought by the Harry Fox Agency against the Universal Music Group of Vivendi Universal S.A. over Universal's online music distribution service, "Farmclub.com") and at the U.S. Copyright Office.

The music publishers also took action against smaller-scale online users of published music scores. For example, the Harry Fox Agency contacted hundreds of Web site operators who were using the Internet to distribute the lyrics or printed musical scores for compositions managed by the agency. Music publishers take the position that online publication of lyrics or music scores falls within the music publishing rights controlled by the agency; thus, such online publication requires a royalty to the agency on behalf of the composers. Online publication of music scores is a form of publishing and requires a publication license. Online distribution of recorded music also requires some form of compensation to the holders of the publishing rights associated with the recorded music that is distributed.

Another example of the complex set of relationships in the world of online music distribution is provided by the conflicts that are developing among the music recording companies, the music performance rights organizations, and music broadcaster (both radio broadcasters and Webcasters). Broadcasters have traditionally paid royalties to the music performance rights organizations for the right to "perform" (i.e., play) music sound recordings on the radio. This means that the broadcasters pay the composers of the music they perform for the right to broadcast that music. Traditional broadcasters do not, however, pay the recording companies or the artists who create the recordings that the broadcasters perform, since U.S. copyright law exempts those performances from royalty payments. The leading performance rights organizations in the United States are Broadcast Music, Inc. (BMI), the American Society of Composers, Authors, and Publishers (ASCAP), and Sesac, Inc. Acting on behalf of music composers and recording companies, the performance rights organizations enter into licensing agreements with the various parties who want to perform music publicly (e.g., radio and television broadcasters, motion picture producers, concert promoters). The parties who perform the music thus pay the performance rights organizations, who in turn compensate the appropriate composers.

A major issue confronting online music broadcasts is the royalty to be paid by Webcasters to artists and recording companies. As noted previously, traditional broadcasters were required to pay royalties to composers (commonly collected by the performance rights organizations); however, U.S. copyright law did not require those broadcasters to pay royalties to the record companies or the artists. The Digital Millennium Copyright Act (DMCA) removed that exemption as to music performances on the Internet, and it required the U.S. Copyright Office to establish an appropriate royalty system for the music Webcasts. Accordingly, under the DMCA, Webcasts of recorded music require a payment from the Webcaster to the artists and record companies, while traditional music broadcasts continue to be exempt from those payments. Not surprisingly, Webcasters are generally troubled by the fact that they are required to pay a royalty that is not required of traditional broadcasters.

Traditional broadcasters argued that they should not be required to pay the new royalty, even if they make some of their conventional broadcast content available online, contending that the royalty was aimed only at pure Webcasters. In its initial action on this issue, the Copyright Office concluded that the traditional broadcasters should be subject to the new royalty for online distribution. The National Association of Broadcasters appealed that decision in federal court. The trial court upheld the determination made by the Copyright Office (*Bonneville International Corp. v. Register of Copyrights*). The broadcasters appealed the trial court's ruling, and that appeal is pending. In spite of the appeal, the Copyright Office continued action on the arbitration proceeding to establish a specific online royalty structure, which is discussed in greater detail in the following text.

The U.S. Copyright Office moved to establish the appropriate Webcast royalty rate in an arbitration proceeding. In its arbitration proceeding, the Copyright Office determined that there should be different royalty rates assessed against the already licensed broadcasters and the Webcaster. The Copyright Office proposed that a fee of 0.07 of a cent per streamed song, per listener, should be assessed for online music that is an online retransmission of broadcast material. The Copyright Office proposed that a fee of 0.14 of a cent per streamed song, per listener, should apply to all Webcasts of music that are not retransmissions of broadcast music. The proposal by the Copyright Office also included establishment of a lower, but as yet unspecified, royalty rate for noncommercial Webcasters. The Librarian of Congress rejected those proposed rates and set an equalized rate of 0.07.

Record companies and Webcasters have widely divergent views as to an appropriate royalty level. The record companies argue that the royalty should be set at a level substantially higher than that proposed by the arbitration panel. Webcasters contend that the proposed rates would drive many Webcasters out of business, and they propose a different royalty structure. The Webcasters advocate a rate that would be a set percentage of the Webcaster's revenues. This way, larger Webcast operations would pay more than would smaller ones. Advocates of this approach note that it parallels the royalty structure that has long been applied by the music performance rights organizations; thus, they argue that the process can be effectively implemented.

6.3 Digital video rights

The video industry avoided many of the content rights battles that confronted their print and music industry colleagues in the early days of the Internet. In part, the delay in encountering those issues was caused by the relative scarcity of consumer access to broadband capacity adequate to support high-quality digital video content distribution. With time, however, that respite for the video content industry is ending, and thus the relative good fortune of the digital video content industry is rapidly fading. The video industry now faces many of the same difficult rights management issues that the other media industries are already attempting to resolve. The same challenges as to control over content in digital form that the print publishing and music industries have faced for several years are now confronting the television and motion picture industry.

Distributors of television programming, including both broadcast stations and cable networks, are increasingly in conflict with online video content providers. For example, there are several instances when online sites attempting to redistribute television content were confronted by television content producers and traditional television content distributors. A company called iCraveTV, for instance, rebroadcast television programming from certain U.S. television stations via the Internet. Challenged by various television stations and program producers, iCraveTV ultimately agreed to cease operations.

Another online video enterprise, RecordTV, was also challenged by television producers for copyright infringement. RecordTV provided a Web-based system, which enabled end users to access diverse television programming that had been recorded and stored on servers operated by RecordTV. After court challenges, RecordTV entered into a settlement agreement in which it agreed to cease its streaming video operations.

An additional issue associated with online television content is the debate over the ability of digital video recording systems to delete commercial advertising from the recorded programs. Digital recording systems such as Tivo and Replay TV permit users to enhance their ability to record television programming. Video content providers are concerned that such systems make it easier for users to copy and distribute television programming. Broadcasters and advertisers are concerned that features that enable users to eliminate advertisements from the video recordings undermine the effectiveness and value of those advertisements. Under traditional copyright law, fair use principles would likely protect a wide range of personal use activities associated with recorded television programming, including deletion of advertisements and noncommercial, limited redistribution of copies of programming. With the advent of the anticircumvention provision of the Digital Millennium Copyright Act (DMCA), however, it is possible that equipment manufacturers that create and distribute video recording devices that support these functions may face litigation from video content developers and distributors under the terms of the DMCA for providing systems that can facilitate misuse of copyrighted material.

There are signs that the motion picture industry is encountering growing tension with regard to online access to movies. At one level, the industry could make its production and distribution operations dramatically more efficient, and thus less costly, through greater use of digital production and digital distribution systems. Many of the major motion picture production companies are actively involved in developing digital distribution systems, and it appears that this trend is inevitable. Key factors yet to be resolved, however, include the overall cost of converting the motion picture distribution chain to fully digital equipment and the development of content security measures that can effectively and economically manage access to the content.

The digital video industry is seriously concerned that widespread online distribution systems for video content could cause difficulties for the industry paralleling those already encountered by the music industry. Just as the music industry has not yet found a way to make effective use of the potential benefits of online music distribution while simultaneously guarding against the risks of piracy, so too is the video industry uncertain as to how to balance those conflicting trends. Pressure on the video industry to manage the risks and rewards of digital distribution more effectively will likely mount as broadband data network capacity becomes more widely accessible to consumers. This trend will, it seems, parallel the evolution of the online music challenge, which developed into a more serious economic and policy issue when more consumers gained access to higher-capacity online networks.

Another parallel between the video industry experience and that of the print publishing and music distribution industries is the fact that the traditional distribution systems for video content also involve many different types of intermediaries and that the existence of those intermediaries makes resolution of the digital rights sharing process very complicated. The publishing industry must balance digital rights among authors, photographers, publishers, retail sellers, and consumers. The music industry must accommodate digital rights held by composers, performers, record companies, music rights organizations, broadcasters, retailers, and consumers. The digital video industry must balance digital rights of producers, broadcasters, cable systems, program syndicators, video distributors, and consumers. Each of these industry rights balancing efforts will continue to be enormously complex.

6.4 Secure systems for digital content

Providers of digital media content moved aggressively to apply technology to protect their electronic material from unauthorized access or use. Encryption technology is, for example, widely applied to digital media content. Often characterized as "digital rights management" systems, these technologies will play an essential role in all digital media distribution systems and strategies in the future; however, they have also generated an additional level of legal controversy.

Several court disputes were initiated as a result of efforts to circumvent digital content security systems. As discussed in greater detail in Chapter 9, the Digital Millennium Copyright Act restricts the development, manufacture, distribution, and use of technologies that can be applied to defeat copyright protection systems. For instance, computer software that enables unauthorized users to access encrypted copyrighted material can fall within the prohibitions of the DMCA. This aspect of the DMCA is commonly referred to as the "anticircumvention provision" of the act.

Some of the most visible disputes involving the anticircumvention provision of the DMCA are the cases involving the "DeCSS" computer program used to decrypt DVD content. The primary case in this dispute is *Universal Studios v. Reimerdes*. In that case, a federal court was asked to apply the DMCA's anticircumvention provision to block the online, noncommercial distribution of the DeCSS program. The trial court issued an injunction barring the online distribution of DeCSS. That decision was appealed, and, after review by the appellate court, the injunction was affirmed. As a result, online posting of the DeCSS program was prohibited, pending final resolution of the merits of the challenge to the constitutionality of the DMCA's anticircumvention provision.

Another example of the legal controversy triggered in the United States by application of the DMCA to restrict the development and distribution of technology with the potential to circumvent copyright protection measures was the case of *Felten v. R.I.A.A.* In that case, a computer science professor and his colleagues asked a federal court in the United States to declare that application of the DMCA's anticircumvention provision to prevent academic publication of a system that could override copyright protection technology was unconstitutional. Professor Felten and his colleagues developed a system that could circumvent new technology under development by the music recording industry for protection of security for digital format music. Professor Felten and his colleagues developed the system in response to a general challenge issued by the music industry, and they intended to publish their findings at an academic conference. Prior to the publication, the music industry took the position that even the academic publication of the findings would violate the DMCA. Professor Felten's claim was rejected by a federal court in New Jersey, and Professor Felten elected not to appeal the dismissal of the case, since the music industry provided assurances that it would not attempt to block such academic publications in the future.

Another federal court case in the United States also speaks to the issue of application of the DMCA against new circumvention technologies. The U.S. government initiated a criminal prosecution against Russian software developer Dimitry Sklyarov under the DMCA. The prosecution is based on Sklyarov's development of software that can be used to circumvent encryption capability applied to the content of certain electronic books. It is the first formal criminal prosecution in the United States brought under the anticircumvention provision of the DMCA. The decision by the U.S. government to pursue criminal prosecution under the DMCA in this case suggests that the government is prepared to make use of its criminal law enforcement authority to send a strong message as to digital media ownership rights.

There is concern that the DMCA's anticircumvention provision can impede technological innovation. Critics of the provision suggest that it can significantly interfere with reverse-engineering activities that were lawful in the past. As discussed previously, reverse-engineering consists of the deconstruction of a technology to understand how it is structured and how it operates. The process of reverse-engineering is legitimate and lawful, provided that it does not involve theft of trade secrets, violation of contract obligations, or direct copying of copyright protected material. If the anticircumvention provision of the DMCA is broadly interpreted, that provision could be used to block reverse-engineering in instances where the technology to be analyzed is involved in the protection of copyrighted material. For example, as noted previously, reverse-engineering of the DeCSS decryption program for DVD content has been challenged as a violation of the anticircumvention provision, based, in part, on an argument that the reverse-engineering of the program provides a base for future circumvention of the program. If that argument is accepted, it could be used as the basis for prohibition of all reverse-engineering of DeCSS, regardless of the purpose for the reverse-engineering.

The anticircumvention provision of the DMCA will play a major role in future efforts to manage digital media rights. Encryption and other security measures are integral parts of any digital rights management system. If the DMCA continues to be aggressively and broadly enforced by the authorities, a clear advantage will be provided to digital content owners relative to content users. Recognize, however, that there are serious concerns and objections raised against such expansive application of the DMCA; thus, its future reach is not yet certain. Prudence suggests that content developers and users alike should monitor the continuing evolution of the

DMCA with substantial care to ensure that they remain fully aware of both their rights and obligations under that statute.

6.5 Role of ISPs and other content distributors

Efforts to manage digital media rights in the online environment also raise challenges for the various intermediaries involved in the online distribution of digital media content. Previously in this chapter we discussed the challenges facing some of those intermediaries (e.g., Webcasters), and in Chapter 7 we will discuss the challenges facing the various "peer-to-peer" content-sharing intermediaries (e.g., Napster). We should also note, however, that Internet service providers and other online service operators (e.g., online auction operators) now commonly confront issues associated with management of rights of access and use associated with the media content of other parties that is distributed using their systems.

The DMCA has been interpreted to give important relief to ISPs and other online service providers as to responsibility for the security of media content of other parties that is distributed via their services. The DMCA requires online service providers to implement remedial practices and processes to be applied in the event of allegations of unauthorized use of copyrighted material on their systems. These so-called "notice and take-down" procedures provide a means through which owners of copyrighted material can obtain relief from the online service provider if their content is misused on the system. Online service providers who provide and effectively implement these notice and take-down procedures are protected from liability for misuse of copyrighted content by users of their services. The DMCA thus places a responsibility upon online service providers to act responsibly and promptly with regard to media content of other parties; however, it shields the service providers who comply with those obligations from legal liability for the copyright law violations of the service users.

More specifically, the notice and take-down requirements of the DMCA place an obligation on ISPs and other online service providers to implement at least the following measures to enforce copyright ownership controls. The service providers must designate an agent and provide contact information for that agent so that copyright owners know who to inform if they believe that some of their copyrighted material has been misused on the online service. The contact information should include a description of the specific information the service provider requires from the copyright owner in order to investigate the claim. The service providers should provide notice to their users that service will be terminated if the users misuse copyright protected material, and the notice should indicate whether or not the service provider intends to take the allegedly infringing material off the service immediately, pending completion of the investigation of the claim. Service providers should make sure that the operations of their systems accommodate and do not interfere with copyright identification and protection measures. Finally, the service providers should provide their users with access to basic information regarding copyright law obligations (e.g., links from their site to online resources that provide basic information regarding copyrights).

When a service provider receives notice of alleged copyright infringement, it should promptly review the allegation. The service provider should also promptly provide notice of the claim to the customers affected by the claim. Those customers should be given an opportunity to respond to the claim. If the service provider concludes that the allegation appears to be accurate, the infringing material should be removed from the system. If the allegation appears to be inaccurate, the service provider should advise both the party raising the complaint and the affected user of the service of the determination made by the service provider.

In the case of *Ellison v. Robertson*, in federal court in the United States, the court provided guidance as to online service provider liability for copyright infringing content under the DMCA. The court dismissed the claims against America Online that were part of the complaint. The court concluded that the DMCA shielded online service providers from liability when their customers posted copyright infringing material in Usenet groups, if the service provider effectively complied with the notice and take-down requirements of the DMCA. To date, courts have been more willing to apply the DMCA liability shield to ISPs than to other forms of online service providers (e.g., file-sharing systems, online auction operators). The *Ellison* case leaves open the door for broader application of the liability shield for online service providers in addition to ISPs in the future.

Service providers should be aware that the DMCA's liability shield applies only to the extent that the copyright infringing conduct by the service user was not supported or encouraged by the Internet service provider. For example, if a service provider compensated the user who engaged in the infringement, or if the service provider gave the infringing content to the end user, the service provider would could lose its protection from liability. The DMCA does not insulate online service providers from copyright liability if they contributed to the infringement engaged in by their customer in some manner beyond merely providing Internet service to that customer.

Another case involving management of digital media rights by intermediaries in the digital distribution chain is the litigation between News Corp. and Vivendi Universal S.A. In that case, Vivendi Universal alleges that News Corp. contributed to the counterfeiting of smart cards that control access to the digital television signals of Vivendi's Canal Plus Group. The case was filed in federal court in the United States and it includes claims that News Corp. violated U.S. copyright laws by breaking the security codes in the smart cards and by making those codes available to counterfeiters, who, in turn, allegedly distributed the codes on the Internet. This case illustrates the value that digital media businesses place on their ability to manage access to their content, and it provides an example of a form of litigation among commercial competitors that may become more common as the struggle over digital media rights becomes more severe.

6.6 Criminal penalties for violation of digital media rights

An important aspect of the increased attention directed toward effective management of digital media rights is the application of criminal law penalties to misuse of copyrighted digital media content. Traditionally, civil remedies (e.g., monetary damage awards and court orders) were the most common legal response to violations of rights of ownership for media content. At present, governments are increasingly willing to apply criminal law sanctions to protect media content ownership interests. As noted previously, for example, the DMCA in the United States provides for criminal penalties, in addition to civil relief, for violations, and those penalties were invoked for the first time in the case of Dimitry Sklyarov. An increasing number of additional examples of application of criminal sanctions for misuse of copyrighted digital content are also available.

In the *News Corp.* dispute with Vivendi's Canal Plus Group, for instance, the copyright piracy claims have been supplemented by claims under U.S. antiracketeering laws, which carry the potential for criminal penalties. Those laws prohibit commercial activities that represent a continuing course of criminal conduct, and the copyright piracy allegations in that lawsuit are presented by the plaintiffs as a course of criminal conduct that justifies the racketeering claims.

Another example of efforts to apply criminal penalties to digital media disputes is provided by the cases brought by U.S. federal law enforcement authorities against groups alleged to be recreational software pirates. In cases such as *U.S. v. Rothberg*, for instance, federal authorities have prosecuted individual software pirates who were not alleged to be involved in piracy for commercial profit. These claims have been brought under the U.S. law known as the No Electronic Theft Act, which provides for criminal sanctions (including fines and prison terms of several years) for theft of copyright protected material that has a market value of more than $2,500. In the past, civil lawsuits for copyright infringement were easily won against small-scale, noncommercial pirates; however, courts generally did not award significant damages, since the economic harm to the copyright owner resulting from recreational piracy was negligible. In order to send a more vigorous message to copyright pirates, however, federal authorities made changes to U.S. copyright law to permit criminal prosecution for even relatively small-scale piracy. The apparently increasing willingness of law enforcement authorities to initiate criminal prosecutions in cases involving piracy of material with relatively small economic value changes the dynamic of digital media content access. It is likely that this trend will continue into the future.

Note that this trend appears to be taking hold outside of the United States, as well. Although the United States has clearly been the most active country regarding expansion of application of criminal penalties to digital media content violations, other countries now appear to be following that path. For example, the European Community now provides for criminal penalties for certain intellectual property rights violation under the terms of the Cybercrime Treaty it is implementing. In the future, other jurisdictions may well adopt a similar approach.

6.7 Managing digital media rights

Currently, effective management of digital media rights presents a major challenge for the developers, distributors, and users of digital media content. We should expect that challenge to increase in the future. Developers are eager to profit from expanded distribution of their material; however, new technologies make it more difficult for them to exercise their ownership rights over the material. Distributors have new methods to facilitate rapid, efficient, and widespread access to high-quality versions of digital media content; yet, they are unsure about the most appropriate economic models to apply to that access. Consumers of digital media have access to technology and services that facilitate access to media content and enhance their ability to exercise choice as to the content they obtain, yet they also face new legal impediments (e.g., the DMCA) restraining their ability to apply the new technology and services.

Developers of digital media content rely increasingly on protective measures to help them control access to their content. Some of those measures are technological (e.g., encryption systems) and others are provided by law (e.g., enforcement of the DMCA). For the foreseeable future, expect content developers to increase their efforts to apply both technical and legal approaches to manage access to their content and to enforce their ownership interest in that content. Digital media owners have strong incentives to act aggressively to extend and to enforce their ownership rights. One important word of caution for digital content owners should also be expressed. Developers of digital media material should avoid overplaying their ownership claims.

Today, there is substantial sympathy at large for the arguments raised by media content developers that suggest that their ability to enforce their legal rights has been undermined by technological advance. If, however, those developers are perceived, over time, by the public and by government authorities, to be exploiting their property in an unfair and unreasonable manner, there could be a policy backlash. For example, if the music industry, while enforcing its legitimate rights of ownership, is ultimately perceived to be restricting content access unfairly or extracting unreasonable monopoly fees for use of its property, legislators, regulators, and courts may become far less willing to support enforcement of content controls. Governments and a substantial portion of the consuming public will support reasonable conduct by digital content owners to enforce their ownership rights; however, those parties will also expect some important benefits for the public. Chief among those expected benefits will likely be lower prices for the content, a wider range of consumer choice as to the content and its use, and a more convenient consumer experience with regard to content access. If those consumer benefits are slow to develop, there is likely to be significantly less sympathy for the future ownership claims of the content providers.

Distributors of digital media content receive helpful legal protection as to their liability for online media content (e.g., the DMCA notice and take-down provisions); however, they continue to face significant uncertainty as to the appropriate economic models to apply to their role as intermediaries in the electronic media distribution infrastructure. Authorities appear to recognize the importance of these intermediaries in the digital media distribution chain, and it is likely that the government authorities will continue to structure and enforce laws and policies in ways that enable the intermediaries to operate effectively in the increasingly digital commercial marketplace. Digital media intermediaries should also behave cautiously to avoid a policy backlash. If the intermediaries make use of the new technical capabilities to increase consumer choice of media content, to reduce the costs associated with access to that content, and to enhance the quality of the consumer experience, then consumer and government support for those intermediaries will likely continue. If, however, benefits to consumers are slow to materialize, there may well be increasing pressure on authorities to be less sympathetic to the commercial requirements of the digital content intermediaries.

Consumers of digital media content are, in perhaps, the most uncertain position of the three key groups of players in the digital media marketplace. Although their opportunities to obtain access to diverse media content are dramatically increasing, their legal rights with regard to that content remain uncertain. In some instances, traditional legal rights of access (e.g., fair use under copyright law) seem to be less reliable than they have been in the past. In addition, some new legal concepts (e.g., the DMCA's anticircumvention provision and use of criminal law penalties for copyright violations) appear to limit the rights of digital media consumers. It appears to be a classic good news, bad news setting for digital media consumers. The good news is that choices and opportunities are expanding rapidly. The bad news is that content suppliers and government authorities now expect consumers to abide by a higher standard of appropriate conduct when it comes to their use of digital media and that consumers have not yet seen the full range of benefits (e.g., lower prices, greater convenience) that providers of new media have long promised.

It is clear that the authorities will continue to expect media consumers to respect and abide by the legal rights of ownership of content providers. Current laws providing those ownership rights will continue to be enforced energetically, and the trend toward expanding the scope of enforceable legal rights associated with digital media content is likely to extend well into the future. It is also likely, however, that consumers will begin to assert other legal rights, enabling them to preserve access to digital content subject to reasonable prices and terms. In addition to rights such as fair use and public domain that are integral parts of traditional intellectual property access principles, expect consumers to make use of other legal rights, such as consumer protection laws and antitrust/competition law principles to help them ensure that they continue to have reasonable access to digital media content. Government authorities are likely to support, and even encourage, this expanded concept of the legal rights available to consumers to sustain effective digital media access by the public.

Selected bibliography

Arista Records, Inc., et al. v. Xact Radio, LLC, Case No. 01 CV 5090 (S.D. N.Y. filed June 8, 2001), Complaint at <www.riaa.com/pdf/xactcomplaint.pdf>.

ASCAP Web site, <www.ascap.com>.

BMI Web site, <www.bmi.com>.

Bonneville International Corp. v. Register of Copyrights, 153 F. Supp. 2d 763 (E.D. Pa. 2001).

Ellison, et al. v. Robertson, et al. (C.D. Cal. March 13, 2002).

Felten, et al. v. RIAA, et al., Case No. CV-01-2669 (D. N.J. 2002), Complaint at <www.eff.org/Legal/Cases/Felten_v_RIAA/20010606_eff_complaint.html>.

Group Canal S.A., et al. v. NDS Group PLC, et al., Case No. C02-01178 (N.D. Cal. filed March 11, 2001), Complaint at <www.actionone.com/clients/actionmarketing/ The_Complaint.pdf>.

Mathews, A. W., "Royalty Fight Threatens Record Industry's Plans to Deliver Songs Online," *Wall Street Journal,* May 1, 2001, p. A1.

Mathews, A. W., "U.S. Arbitration Panel Proposes Royalty Rate for Webcast Songs," *Wall Street Journal,* Feb. 21, 2002, p. B6.

Mathews, A. W., and B. Orwall, "Major Studios Discuss Plan to Equip Theaters to Show Digital Films," *Wall Street Journal,* May 17, 2001, p. A1.

MGM, et al. v. RecordTV.com, Complaint and Settlement Agreement at <www.mpaa.org/ Press>.

National Football League, et al. v. TVRadio Now, et al., Case No. 00-120 (W.D. Pa. 2000), Complaint and Settlement Agreement at <www.mpaa.org/Press>.

New York Times v. Tasini, Case No. 00-201 (U.S. Supreme Court 2001).

Random House, Inc. v. Rosetta Books, LLC, 150 F. Supp. 2d 613 (S.D. N.Y. 2001).

Rich, L., and H.C. Lee, "Hollywood: Beware of Napsterization," *The Industry Standard,* July 31, 2000, p. 88.

Sony Music Entertainment, et al. v. MTVi Group, LLC, et al. (S.D. N.Y. filed June 8, 2001), Complaint at <www.riaa.com/pdf/mtvicomplaintfiled.pdf>.

The Rodgers & Hammerstein Organization, et al. v. UMG Recordings, Inc., et al., Case No. 00 Civ. 5444 (S.D. N.Y. 2001).

Twentieth Century Fox Film Corp., et al. v. iCraveTV, et al., Case No. 00-121 (W.D. Pa. 2000), Complaint and Settlement Agreement at <www.mpaa.org/Press>.

United States v. Rothberg, et al., Case No. 00CR 85 (N.D. Ill. 2001).

United States v. Sklyarov, Case No. 5-01-257 (N.D. Cal. 2001), Complaint at <http://cryptome.org/usa-v-sklyarov.htm>.

Universal Studios, et al. v. Reimerdes, Docket No. 00-9185 (2d Cir. 2001), Opinion at <www.2600.com/news/112801-files/universal.html>.

Zeran v. America Online, 958 F. Supp. 1124 (E.D. Va.), affirmed, 129 F.3d 327 (4th Cir. 1997), cert.denied, 118 S. Ct. 2341 (1998).

Zomba Recording Corp., et al. v. MusicMatch, Inc., Case No. 01 CV 5091 (S.D. N.Y. Filed June 8, 2001), Complaint at <www.riaa.com/pdf/musicmatchcomplaint.pdf>.

7

The challenge of peer-to-peer and distributed computing

In the previous chapter, we discussed the benefits and conflicts associated with the distribution of traditional media content in digital format. One of the key technical means supporting the expanded use of new media is the digital file-sharing network. Increasingly widespread use of peer-to-peer (P2P) content-sharing systems and distributed computing networks dramatically improves the efficiency and quality of digital media distribution. However, those systems also raise important intellectual asset management challenges. Peer-to-peer systems facilitate rapid access to digital content by large groups of individual users. Distributed computing networks enable numerous separate computers to function collectively, as one, enhancing the efficiency of computer resource use. These systems foster the decentralization of content access and control. While empowering end users, they can make management of rights of access and use a significantly more difficult task. Yet, these same systems can also be applied to management of network content and operations. File-swapping systems and distributed computing networks thus cause major challenges for digital content owners, and they are at the same time part of the solution to those content management issues. In this chapter, we focus on the opportunities and challenges that these content- and resource-sharing systems present for owners, distributors, and users of digital content.

7.1 The P2P process and its challenge

Peer-to-peer systems enable computers that are running a common file-sharing program to facilitate content access quickly and easily. Files stored on one computer on the P2P network are essentially accessible to users of all of the computers on the network. In this way, a virtual library is created, enabling the computer users to increase dramatically the digital content accessible to them. At a basic level, P2P systems create an environment in which the content of one computer on the network is, in effect, content accessible to all computers on the network. Initially, P2P systems required use of an intermediary server to process the file-search and file-sharing functions. The second generation of P2P systems no longer requires use of such intermediaries, supporting instead direct file sharing among the client computers on a network.

The user convenience and flexibility that P2P systems provide also creates some challenges for content developers, system operators, and end users. Content developers fear that P2P systems lead to a loss of control over digital rights management. System operators fear that they will face liability for the conduct of system users, and they are uncertain as to their obligations to enforce the rights of the content owners. System users fear that they will be the targets of legal action by content owners enforcing their rights as to use of digital property. The framework of content ownership and access rights associated with digital material on file-sharing systems remains uncertain, at present.

To date, the greatest attention has been focused on the impact of P2P systems on music distribution. In the future, however, all forms of digital media content will face the same challenge from content-sharing systems that the music industry confronts today. The clear trend is toward an ever wider range of end users sharing an ever diversifying suite of content using file-sharing networks. In that environment, content developers, owners, distributors, and users must all understand the scope of the content-sharing operations. All of those interested parties must work together to devise principles of ownership, access, and economic compensation that preserve the essential incentives necessary to keep diverse digital media content accessible on P2P networks. Those necessary principles may prove to be significantly different from the ones traditionally applied to the distribution of media content.

7.2 Managing Napster

The most highly visible debate regarding control of digital content on P2P networks involved the online music sharing system, Napster. The Napster system was a first generation P2P system; thus, it made use of an intermediary server to process the file sharing. In a series of court cases, composers, musicians, and recording companies all sued Napster for copyright law violations based on the online music file sharing operated by Napster. Among other claims, the plaintiffs contended that the Napster system contributed to copyright infringement on the part of Napster system users and that it constituted a form of copyright protection circumvention that violated the anticircumvention provision of the Digital Millennium Copyright Act (DMCA). Napster's operators were accused of contributory infringement, based on the allegation that their system contributed to the direct infringement engaged in by the end users who were sharing the copyrighted music files. Contributory infringement exists when a party facilitates or enables direct copyright infringement by others. In the *Napster* case, copyright owners contended that Napster provided the technical system that enabled end users to share music files in ways that constituted direct infringement; thus, they claimed that Napster was a contributory infringer. In addition, copyright owners claimed that Napster should be held vicariously liable for the direct infringement of the end users, since Napster derived economic gain as a result of the infringing conduct. Liability for vicarious infringement exists when a party derives economic gain from the direct infringement of another party. The anticircumvention claim was based on the contention that the Napster system constituted, in effect, a technology that could be used to defeat copyright protection measures established by the music copyright owners.

The federal district court considering the case concluded that the Napster operation, as initially structured, presented enough of an infringement threat to justify issuance of a temporary injunction ordering Napster to block distribution of all music files that had not been properly licensed from the copyright owners, pending resolution of the merits of the case. The appellate court that was asked to review the trial court's decision to issue the temporary injunction affirmed the trial court's conclusions. The underlying copyright infringement case between Napster and the record companies remains active and is in the discovery phase. Napster agreed to sell Bertlesmann AG all of the Napster assets and to reorganize the company under Chapter 11 of the U.S. Bankruptcy Code; however, the Bankruptcy court rejected the sale. At the same time, the recording companies have also launched their versions of online music distribution systems in an effort to compete with the P2P systems. For example, EMI, BMG, and Time Warner launched their MusicNet online service. Sony and Universal launched their online music distribution service, Pressplay. The music distribution systems launched by the record companies are structured to require payment for access to the online music. Some observers contend, however, that the online music distribution services offered by the record companies have significantly fewer recordings available than do the P2P systems.

In litigation, Napster attempted to defend itself, in part, by arguing that it had not directly violated any copyright rights, since it had not directly duplicated or distributed the copyrighted material of other parties. Napster claimed that the actual file-sharing activities were performed by the end users. This argument was not sufficient to protect Napster from responsibility, however, as the court concluded that Napster contributed to copyright infringement and that its technology facilitated circumvention of digital rights management systems applied to recorded music. This theory of liability for the operators of file-sharing systems and for the developers of file-sharing technology has significant implications for developers, operators, and users of P2P systems. Napster's legal problems demonstrate that under current interpretation of copyright law, all of the parties involved in the file-sharing distribution chain can be held legally accountable for any misuse of copyrighted material that takes place in that distribution chain.

Napster also argued that the type of content sharing in which its users engaged was private, noncommercial sharing. The Napster legal theory characterized this noncommercial sharing of copyrighted material for personal use as fair use. The court did not, however, accept this line of reasoning either, and the fact that the court did not apply the fair use defense to the Napster process also has important implications for other file-sharing systems. The action by the Napster court, with regard to the fair use argument, suggests that the concept of fair use may be interpreted differently in a file-sharing context than it has been when the material in question was distributed through transfer of ownership of a tangible good (e.g., book, CD, DVD). Fair use has traditionally been interpreted to provide a defense for limited noncommercial duplication and distribution of copyrighted works. It is unclear the extent to which the fair use exemption continues to have meaning in the file-swapping environment. A legitimate purchaser of copyrighted material has traditionally been permitted to make a limited number of copies of that material for noncommercial use by friends an family. It appears, however, that such limited fair use duplication and sharing may be much more difficult to enforce when the material is introduced into a P2P network. The question of what constitutes limited noncommercial fair use in a file-sharing network remains unresolved.

Another basic tenet of copyright law that is challenged by the expansion of P2P networks is the concept of first sale rights. The first sale rule is a doctrine of copyright law. It grants to the buyer of a copy of copyrighted material broad rights of use for the individual copy of the work purchased by the buyer. The buyer traditionally obtained the right to sell or lease the purchased copy of the work. The buyer also traditionally obtained the right to make a limited number of copies of the work for private, noncommercial distribution. The principle of these first sale rights associated with copyright protected material is called into question when the material is introduced into a P2P network. The extent to which first sale rights continue to exist in a file-sharing system remains uncertain at present. It is unclear whether the concept of limited distribution rights for personal use continues to be feasible in a file-swapping environment, where a single copy of a work can be accessed by large numbers of individual users who participate in the network.

One of the lessons derived from the court's treatment of Napster is the fact that operators of file-sharing systems will be held accountable for misuse of copyright protected material on their systems. Material for which copyrights are not being enforced can be safely distributed on P2P networks. Copyright protected material must, however, be guarded by those networks. Operators of the networks will be held responsible for unauthorized use of copyrighted material, even if the system gives end users the substantial control over the material they access. As a result, providers of file-sharing services must either pay royalties to the owners of copyrighted material distributed using their systems, or they must impose operational safeguards to block the sharing of unlicensed copyrighted content.

Operators of file-sharing networks must adopt notice and take-down procedures of the sort contemplated for online service providers by the

DMCA. As noted in Chapter 6, the DMCA's notice and take-down requirement forces ISPs and other online service providers to implement policies and procedures to ensure that material on their systems that infringes on copyright ownership rights of other parties will be taken off the system promptly when the infringement is brought to the attention of the service provider. The Napster example illustrates that courts appear to be willing to treat file-sharing network operators as online service providers and to hold those service providers to the standards applied by the DMCA. Recognize that the universe of online service providers and ISPs that fall within the scope of the DMCA is large. Operators of institutional networks that have Internet access (e.g., universities, commercial businesses) also fall within the scope of the DMCA and should also abide by the notice and take-down process.

Internet service providers are not defined as telecommunications common carriers. In the *Napster* case, Napster tried to argue, in part, that it was, in effect, a common carrier; however, the court rejected that argument. Traditional telephone companies and other common carriers received exemption from liability for the nature of the content transmitted by the users of their systems. For example, telephone companies are not liable for copyright infringing material or defamatory material that is communicated using their systems. In exchange for that immunity from liability, the telecommunications common carriers were subject to federal regulation of their rates and terms of service. Internet service providers successfully managed to avoid being classified as common carriers, thus their rates and terms of service are not regulated; however, they are generally not immune from content liability. Courts have tended to treat ISPs in much the same way that they handle print publishers (e.g., newspapers, magazines). Some limited exemptions from content liability have been provided for ISPs, however. As noted previously, ISPs that abide by the notice and take-down requirements of the DMCA are protected from liability for the copyright infringement of their customers. Another example is the exemption of liability provided by the Telecommunications Act of 1996 for ISPs that unknowingly serve as conduits for content that is harmful to minors.

Recognize that widely used file-sharing networks can present copyright liability issues for several different network operators. In the *Napster* dispute, the primary P2P operator that attracted the attention of copyright owners was Napster itself. The content providers directed their legal claims against Napster and did not pursue the individual Napster users or the operators of other networks used by end users to access Napster. For example, many college and university computer networks were used as gateways to reach Napster, since Napster's music content was particularly popular among students. Although the music industry did not choose to take such action, legal claims under contributory or vicarious copyright liability theories could have been pursued against the colleges and universities that operated the networks used by the students to access the infringing content. The owners and operators of computer networks and online services that enable individual users to participate in P2P and other file-swapping systems face potential copyright liability, even if they are not the providers of the file-sharing software. It is likely that copyright owners will be increasingly willing to take legal action against these network and service operators, as the file-sharing systems become more decentralized and it becomes more difficult to identify a single entity that is truly managing the content-sharing system. When file-sharing systems become less reliant on a single gateway or intermediary at the center of the content-sharing process, copyright owners will have greater incentive to take legal action against the various network operators and online service providers that enable the end users to join the file-sharing networks.

End users of P2P systems also face copyright infringement liability. In the *Napster* case, copyright owners elected to pursue their legal claims against Napster, not the individual Napster users. That decision was a tactical one; however, copyright owners continue to have the ability to sue individual P2P users for infringement, and it is likely that we will see an increasing number of such legal suits in the future. Copyright owners chose to sue Napster instead of the individual users, since they viewed that line of attack as the one most likely to stop the unauthorized distribution most quickly. Lawsuits against end users will be the cases that will most clearly raise and address the issue of how to balance traditional end users' rights of fair use and first sale rights against the ownership interests of copyright holders. One of the primary challenges associated with file-sharing systems is that they blur the distinction between personal ownership and use of content and widespread redistribution of that content. When end users are linked in ways that make much of their content accessible to many other individuals, personal use becomes, in effect, community use. That setting is substantially different from the one originally contemplated when basic fair use and first sale concepts developed.

Fair use and first sale concepts of copyright law have been tested by new technologies and innovative business models in the past. For example, the advent of videocassette recorders dramatically challenged fair use and first sale principles. Commercial distribution of VCRs led to litigation between the motion picture and television industries and VCR manufacturers (*Sony Corp. v. Universal City Studios, Inc., et al.*). In that case, the U.S. Supreme Court ultimately concluded that the private, noncommercial use of VCRs to permit recording of video content for more convenient viewing was not a violation of copyright law and constituted fair use. As the VCR became more commonly used, video rental operations (e.g., Blockbuster) came into existence, and those distributors were initially opposed by video content producers. Eventually, the video rental operators negotiated contractual relationships with video producers and become an important part of the video distribution chain. Although they were originally vigorously opposed by television and motion picture producers, video rental operators eventually became an important source of licensing revenue for those producers. P2P technologies and other forms of content sharing also probe the boundaries of traditional fair use and first sale doctrines, and those doctrines should be carefully applied to accommodate the capabilities of the new technologies.

The Napster controversy has generated criticism both for content users and for content owners. Some critics of Napster and other content-sharing systems view them to be little more than copyright piracy havens, networks that make it easy for individual users to steal copyrighted works. Other observers are critical of the highly litigious approach that the music industry adopted in response to the rise of the P2P systems. Those observers suggest that P2P networks will be among the most important media distribution systems in the future and that media content owners should not rely exclusively on legal action to respond to the content-sharing process. Instead, many of those critics of the music industry argue that the industry should move more quickly to integrate P2P systems into the music distribution chain, using the new system to help reduce the cost associated with music distribution and to enhance the responsiveness of the music industry to consumer demands. It is likely that, for the foreseeable future, the music industry (and the other traditional media industries, as well) will hedge its bets, applying both legal actions to attempt to slow the expansion of the P2P network and affirmative efforts to incorporate P2P technology into their distribution chains. We are beginning to see early efforts by copyright owners to integrate P2P technologies into their distribution chains through music distribution systems such as Pressplay and MusicNet, for example, and we can expect those efforts to continue into the future.

7.3 The DMCA as a threat to P2P systems

As noted previously, the Digital Millennium Copyright Act's anticircumvention provision has been used as an effective weapon against P2P systems. The DMCA is a particularly potent weapon, since it can be used prospectively. Copyright owners need not wait until they can prove actual economic damages resulting from copyright infringement before they seek legal relief. The anticircumvention provision of the DMCA enables copyright owners to fire a preemptive strike against file-sharing systems and other new technologies. The provision provides for relief based on the potential use of a technology to override measures designed to protect copyrighted material. This approach makes it possible for copyright owners to go to court to stop use of systems that can defeat copyright protection measures, even before those systems are operational. If the anticircumvention provision had been in effect when previous media technology advances occurred—for example, when the videocassette recorder first went into commercial distribution—it is likely that the law would have blocked the development of those technologies.

The provision has already been applied in an effort to delay commercial launch of systems and technologies in addition to Napster. For instance, the DMCA was used by music recording companies in an effort to block commercial release of the first portable MP3 player, the Rio. In that case, the court ultimately concluded that the Rio performed the function of conversion of audio recordings into digital format and, as a result, was a computer peripheral device permitted under the Digital Recording Act. Based, in part, on that analysis, the court permitted the commercial distribution of MP3 players. The DMCA was also invoked in legal challenges against MP3.com, a service that assisted parties who possessed CDs to obtain MP3 versions of the music on their CDs. MP3.com created a stored library of music in MP3 form and delivered copies of the stored music to system users after verifying that they possessed corresponding versions of the requested music in CD form. MP3.com thus argued, in part, that it was merely engaged in music storage and format shifting. The record companies argued, in part, that the entire MP3.com system provided a means of circumventing music copyright controls and was thus a violation of the DMCA. In the *MP3.com* case, the court determined that MP3.com's system violated the DMCA, and MP3.com was required to obtain licenses for the copyrighted music that it delivered.

As this history suggests, there has been substantial pressure exerted by copyright owners to enforce the DMCA expansively. If the anticircumvention provision of the act continues to be applied broadly, it can impede expansion of P2P use. A broad interpretation of the scope of the DMCA's anticircumvention provision can also restrict development of new technologies that are not yet anticipated. Concern about this unintended impact of the DMCA has led some critics of the law to propose amendments that would restrict the scope of the anticircumvention provision.

7.4 The next generation P2P systems

The first generation of P2P systems (including Napster and Scour) used servers to manage the search of share files stored on the various client computers. The next generation of P2P systems was structured so that the systems did not rely on centralized computers to handle the file search function. This next generation of P2P systems is thus substantially less centralized than the first generation. In addition to Napster, a wide range of P2P systems is now popular, and most of those newer P2P systems are more difficult to control than Napster and can process a wider range of digital content. These systems are in use to facilitate sharing of content in addition to music files. Computer software, video, and multimedia materials are now commonly shared, thanks to the many different P2P systems now available. Many different file-sharing systems are now in operation and their popularity continues to grow. Some of the most widely recognized file-sharing systems include Morpheus (MusicCity), KaZaa, iMesh, BearShare, LimeWire, Audiogalaxy Satellite, Gnutella, and Grokster.

Several of these new systems have, like Napster, been sued by copyright owners who claim that the systems are used to circumvent anticopying systems. For example, media content producers have sued Grokster, challenging the KaZaa and Morpheus peer-to-peer systems (*MGM, et al. v. Grokster*). Copyright actions have been brought against Scour (*Twentieth Century Fox, et al. v. Scour*). The Aimster system is the target of litigation as well (In re *Aimster and Zomba v. John Deep*). Expect each new version of the file-sharing process to be challenged in court by traditional media content interests.

The second generation of P2P systems is more decentralized than the first generation. No longer reliant on a server to process search requests, the newer P2P systems provide end users with greater control over the material they access. These systems are also more difficult to monitor, since the content sharing takes place outside of the scope of any centralized authority. As the network becomes less centralized, the challenge of managing access to the content of the network is made more difficult. In these more widely distributed P2P networks, there is often no single organization to be held accountable for the material distributed through the system. In such a setting, copyright owners could be faced with the challenge of trying to enforce their ownership rights against many different individual users in many different jurisdictions. We already see that many of the P2P service providers and end users are located outside of the United States, and that trend is likely to increase in the future, particularly if the DMCA continues to be applied broadly. That environment would pose a complex enforcement challenge for copyright holders, since they may be forced to initiate copyright cases in foreign courts or attempt to have U.S. court judgments enforced by less sympathetic foreign courts.

7.5 Distributed networks

Distributed computing enables network operators to use the collective computational resource of networks more efficiently. Computing capacity not being used by their operators, at a particular time, can be made available to other users who have substantial computing resource needs. A distributed computing process makes computer networks more efficient, more flexible, and more robust; however, it also increases the content management challenges. Since each user of the network has some degree of access to, and control of, the computers of other users of the network, system content is widely shared.

In a distributed computing environment, control over network operations is decentralized. The amount of computing resources accessible to the users of the network increases. This expands the ability of any single network user to access and use the collective content of the network. Distributed networks can promote more diverse use of the shared content than we have yet seen from the various P2P systems that have been widely adopted to date. Although those more diverse content uses can enhance network efficiency and user flexibility, they also expand the number of different applications for all network content that is available for each network user. A large distributed network poses significant digital rights management challenges.

7.6 P2P-driven content searches

An increasingly popular application for P2P systems is to expand the scope of online content search capabilities. Companies such as InfraSearch and Pointera make use of P2P software to drive search engines. This approach enables the engine to access a wider range of stored files, since it can access all of the files on computers participating in the network. Effective search functions are essential to the development of truly useful file-sharing networks. As these P2P search functions become more effective, the value of the underlying P2P network is enhanced.

Expanded and more effective P2P search capabilities also have an impact on digital rights management. At one level, that enhanced capability can make the challenge of content management more difficult, since it will make more material available for sharing and will encourage more users to join the network. At another level, however, more sophisticated search functions can also assist digital rights management efforts. For example, file search capabilities that can more effectively identify and track copyright and license information, including restrictions and limitations on use, can provide valuable support for digital rights management processes. As P2P search capabilities become more sophisticated, they can both undermine and enhance digital rights management activities.

7.7 Distributed computing in P2P systems

There are now examples of P2P systems that support distributed computing functions. For example, the KaZaa Digital Media Desktop P2P system reportedly includes a distributed computing capability. It has been announced that a company called Brilliant Digital Entertainment, Inc., has been granted the right to activate the distributed computing function in the KaZaa product. Brilliant Digital will reportedly use the KaZaa software to make use of the client computers to perform computing functions on behalf of Brilliant Digital's customers. Among the functions to be performed by Brilliant Digital is storage of digital media content. For example, Brilliant Digital will apparently use the distributed computing capability to store digital video programming on the computers of various KaZaa users, instead of storing that content on a single server owned by Brilliant Digital. When the Brilliant Digital customer wants to watch the video, the KaZaa system will search for the files and will download them to the viewer.

This activity has stirred some controversy. Some observers contend that the practice is an intrusion into the privacy rights of KaZaa software users and that the intrusion can pose a security threat to those users. Brilliant Digital and KaZaa argue that the standard licensing agreement accepted by KaZaa users when they download the software expressly grants KaZaa the right to make use of the user's computer through the distributed computing capability built into the KaZaa software. It is believed, however, that a significant number of users of the software do not focus on the rights of use that they grant to the software provider by accepting the software. Expect the debate over this type of application to continue. Parties who intend to make use of this type of combined P2P and distributed computing function should be sure to clearly describe the function to all parties affected by the process, and they should be sure to obtain express indication of consent from all of the users who will be affected by the distributed computing process.

7.8 Content management in P2P and distributed networks

Certain lessons as to how digital content can be managed in P2P settings are emerging. Critical issues to be resolved before digital rights can be effectively managed in content-sharing systems include content security, efficient search mechanisms, and effective user and content authentication processes. It will also be necessary to clarify the scope of owner and user rights of control and access. Finally, effective economic models must also be established to govern use of the shared content.

P2P and other content-sharing systems will likely require adjustments to the traditional content licensing model. For example, greater use of blanket licenses for digital material instead of sales of physical copies (e.g., CDs or DVDs) provides a more effective licensing structure in an environment where content, once distributed, is so readily accessible to multiple users. Content providers may be more comfortable distributing their works under fee structures that tie compensation to the number of users of the material. Those licenses will likely make use of both flat fees (e.g., subscription fees) and small per transaction charges. Licenses tied to the number of users and the term of the license may also be more popular in the content-sharing environment.

Some forms of content sharing and distributed computing technologies are now being applied to manage digital rights more effectively in networked environments. Companies such as Groove Networks, ThinkStream, xDegrees, and Consilient are applying, in various ways, content sharing technologies for use to manage digital rights in commercial settings. These efforts are also in progress at some high-profile, large computer industry companies, including Microsoft (through its .Net initiative), Intel (through its P2P Working Group), and Sun Microsystems (through its Project JXTA). These efforts to apply file-sharing systems in business environments have in their heritage both P2P systems and the various versions of groupware software that have long been used to support collaborative activities in business environments. They generally involve the use of file-sharing systems, for commercial projects, by authorized users.

The key challenge for content owners in the P2P environment is to develop an appropriate pricing model to apply to their goods. Such a model should be structured to ensure that the content developers remain profitable in spite of the inevitable level of piracy that will exist. In addition, that pricing model should reflect the new characteristics of the product being sold. For example, the current pricing model applied by the music recording industry is structured on the concept that the goods being sold are the collections of songs (i.e., CDs, albums). Given the capabilities presented by P2P and other file-sharing systems, a strong argument can be made that, in the future, the product to be sold by the recording industry will be individual songs, not collections of songs. In that more fragmented setting, a different pricing model is appropriate. A similar trend toward pricing based on access to smaller units of content (e.g., individual articles instead of entire collections or video excerpts instead of an entire program) will likely appear in media in addition to music.

Another important issue is the need to clarify the relative rights of ownership and access associated with the content made accessible using file-sharing networks. To date, traditional legal principles of intellectual property ownership (e.g., copyright) have been expanded to help digital content owners to enforce their property interests. Traditional intellectual property law principles providing user access (e.g., fair use, first sale rights) have not yet, however, been clarified in the context of shared digital content networks. In addition, legal principles of general property law (e.g., trespass) and commercial transaction regulation (e.g., the Uniform Computer Information Transactions Act) now have a direct bearing on rights of ownership and use applicable to shared digital content. This means that, in the future, both owners and users of content on file-sharing systems will have an expanded array of legal theories they can apply to establish their rights of access and use.

Also of significance is the fact that technological advances and enhancements will simultaneously increase the difficulty of digital rights management and provide measures to help enforce those rights. For example, the diverse applications being developed for P2P systems make it more difficult for content owners to control access to, and use of, their property. At the same time, however, the expanded search and control capabilities now being integrated into file-sharing systems can help to identify and track ownership and licensing information relevant to the content of those systems. Technology will continue both to challenge and to assist owners, distributors, and users of content-sharing systems.

Content-sharing systems will become increasingly decentralized as time passes. In the future, expect fewer systems that involve a centralized access point (e.g., Napster) and more systems that make use of more widely distributed content access. More decentralized networks make the digital rights management issues more difficult to handle; however, they are likely to be the norm in the future.

A final significant trend relevant to P2P systems is the fact that their popularity will likely continue to increase well into the future. The trend toward increased acceptance of, and reliance upon, content-sharing systems will continue. All parties in the media distribution chain should recognize this trend and accommodate it. Content developers and owners should find ways to incorporate P2P systems into their distribution chains. Content distributors should devise strategies to ensure that they continue to add value to the content distribution process, even though their roles as intermediaries may change as P2P systems become more prevalent. Finally, content users should recognize that the increased control and choice that P2P systems provide for them comes at a price and that the price associated with that empowerment is a greater responsibility to understand the full extent of their legal rights and obligations and a duty to respect the rights of content owners.

Selected bibliography

A&M Records, et al. v. Napster, U.S. App. LEXIS 1941 (Ninth Cir. 2001).

Aimster Web Site, <www.aimster.com>.

Consilient Web Site, <www.consilient.net>.

Freenet Web Site, <http://freenet.sourceforge.net>.

Gnutella Web Site, <http://gnutellawego.com>.

Groove Networks, <www.groove.net>.

iMesh Web Site, <www.imesh.com>.

In re Aimster, Case No. 01 C 8933 (N.D. Ill. 2001).

Jungle Monkey Web Site, <www.junglemonkey.net>.

KaZaa Web Site, <www.kazaa.com>.

Metallica, et al. v. Napster, Case No. 00-13914 (S.D. Cal. filed Apr. 13, 2000).

MGM Studios, et al. v. Grokster Ltd., et al. (C.D. Cal. filed Oct. 2, 2001), Complaint at <www.mpaa.org/Press>.

Mojo Nation Web Site, <www.mojonation.net>.

Morpheus Web Site, <www.musiccity.com>.

Napster Web Site, <www.napster.com>.

Project JXTA Web Site, <www.jxta.org>.

RIAA, et al. v. Napster, Case Nos. 00-16401 and 00-16403 (Ninth Cir. 2001), case histories and pleadings available at <www.riaa.com>.

Scour Web Site, <www.scour.com>.

Sony Corp. v. Universal City Studios, Inc., 464 U.S. 417 (1984).

ThinkStream Web Site, <www.thinkstream.com>.

Twentieth Century Fox Film Corp., et al. v. Scour, Inc. (S.D. N.Y. filed July 20, 2000), Complaint at <www.mpaa.org/Press>.

Wingfield, N., "Brilliant Digital Aims to Harness Power of Users' PC," *Wall Street Journal*, April 3, 2001, p. B10.

XDegrees Web Site, <www.xdegrees.com>.

Zomba Recording Corp., et al. v. John Deep, et al., Case No. 01 CV 4452 (S.D. N.Y. filed May 24, 2001).

8

Property law and commercial transactions law applied to digital content

Effective management of digital content now requires an understanding of some basic concepts of property law and the law of commercial transactions (i.e., contracts and sales). Whether you are a developer or a user of digital content, you will be affected by these legal concepts. Your strategy for management of digital material should thus reflect the rights and obligations now being established for digital content by these laws. Understanding those legal principles in the context of digital content is not a simple task, since property and commercial law concepts in the digital domain are rapidly evolving. This chapter will provide insight into the current landscape of those concepts and some suggestions regarding the future evolution of property and commercial transaction law, as applied to digital material.

8.1 Database ownership

The European Community (EC) has established legal rights of ownership for collections of data. The European Database Directive (Directive 96/9/EC of the European Parliament and the Council of March 11, 1996) established the principle of ownership of aggregations of data, and that principle was enacted by local laws implemented by the various EC member countries. Essentially, the EC database ownership laws grant the creators of data sets and other collections of data the right to require compensation in exchange for rights of access or use of their data. Makers of data collections thus have the legally enforceable right to be compensated by users of their data for the costs of developing and maintaining the data collections. Grant of ownership rights for collections of data fills the traditional gap created by the fact that the intellectual property law principles of copyright protection did not apply to sets of data, since copyright law included an assumption that collections of data lacked the requisite creative component to merit protection.

The EC Directive established the basic legal principles to be applied to ownership rights for the content of data collections. Each of the EC member countries then enacted their own national laws to implement the terms of the Database Directive. For example, the United Kingdom implemented the directive standards through its Copyright and Rights in Databases Regulations (1997). Similarly, the Netherlands enacted the directive with its Database Act (1999), and Spain's enabling legislation took the form of its 1998 Amendment to the IP Act of 1996. Each of the national database laws that implement the directive is interpreted and enforced by the courts in that country.

The European database ownership rights have broad applicability. The rules apply to collections of data in either electronic or paper (hard copy) form. Qualifying material in a computer file or in a print publication is subject to the protection provided by the database ownership laws. A court in Belgium, for example, determined that information regarding self-help groups published in a print pamphlet qualified as a data collection subject to the protection of the database ownership laws (*UNMS v. Belpharama Communication*). A German court concluded that telephone directory listings, in either print or electronic form, qualified for protection as eligible data collections (Tele-Info-CD-Bundesgerichtshof).

Database ownership rights can only be enforced by the maker of the data collection, provided that the maker falls within the jurisdiction of the EC. To be within the scope of the EC jurisdiction, the database maker must be a citizen of one of the EC member countries or a firm that is formed under the laws of an EC country and has a registered office or a principle place of business in an EC country. Thus, an individual person who is a national of an EC country could enforce data ownership rights.

So, too, could a company that was created under the laws of an EC country and that is registered to do business in an EC country also enforce these rights.

The party who actually compiled a data collection is eligible to enforce ownership rights under the directive. There are also indications that some other parties, who may have been involved in the creation of the database but did not directly perform the process of creation, can also enforce the ownership rights. For example, a German court determined that the owner of a database can enforce these rights even if the database was actually developed by a contractor (*C. Net* Case). The court's rationale for this conclusion was that the owner of the database invested substantial resources into the creation of the database (i.e., invested money to pay for its creation) and was thus entitled to the protection afforded by the data ownership rules.

It remains an open issue as to how many different parties can enforce rights under the directive. At present, the creator of the database can certainly enforce those rights. As the German court indicated, it is also possible that a party who pays for the creation of a data collection can enforce database rights. The key element appears to be the extent to which the party claiming rights under the directive has invested substantial and quantifiable resources into the creation and maintenance of the data collection. The basic concept behind the grant of rights in the directive was the desire to provide an enforceable system to enable parties who have legitimately incurred costs associated with creation and updating of data collections to recover those costs from the parties who access or use the material. The directive was generally intended to provide a legal basis for makers of data collections to protect their investment of resources in the collection, verification, presentation, and updating of information. Thus, a party seeking to enforce database rights under the directive must show that it invested significant resources in the creation and maintenance of the database.

Several courts have addressed the issue of what constitutes an investment of resources sufficient to trigger the rights established by the directive. The investment of resources by the database maker must be determined to be substantial before a court will enforce that maker's right to be compensated for use of the data. For instance, a French court concluded that compilation of a list of procurement calls for tenders (bids) did not represent a significant enough investment of resources by the compiler to subject the list to protection under the directive (*Group*

Moniteur v. Observatoire des Marches Publics). A court in the Netherlands determined that the directive did not provide protection for databases that were simply spin-offs from other works (*NOS v. DeTelegraaf*). In that case, the data collection at issue was a listing of television programs created by a television broadcaster. That court's finding that spin-off works were not protected was challenged, however, by one court in the Netherlands (*KPN v. XSO*) and supported by another (*Algemeen Dagblad v. Eureka*).

The issue of precisely what degree of resource investment constitutes a level significant enough to invoke the protections of the directive remains uncertain. Suffice it to say that each court reviewing a claim under the Database Directive must examine the specific facts presented by the maker of the data collection and must determine, based on those facts, whether the maker has invested resources into the creation and maintenance of the database that are sufficient to bring the rights granted under the directive into effect. Makers of data collections who seek to enforce ownership rights under the directive should, however, create a record that documents the level of their resource investment in the creation and management of the collection, since that record will play a critical role in enabling a court to establish the significance of that investment.

The directive permits recovery of costs only to the extent that there has been repeated extraction of material from the data collection or substantial use of the data in the collection. This means that if the content of a data collection has been accessed infrequently, on a sporadic basis, or if the amount of material from the data collection that has been used is insubstantial, the maker of the data collection in question will not be able to recover costs from the user. Obviously, this provision means that a key factual question in every dispute under the directive is whether there has been substantial use or repeated extraction of the content.

Some courts in Europe have begun to interpret the substantial use and repeated extraction requirements of the directive. For example, a German court concluded that repeated extraction of very small amounts of information from a data collection can constitute both the repeated extraction and the substantial use necessary to permit the database maker to recover costs under the directive (Berlin Online). In that case, an online database was searched repeatedly by a metasearch engine that automatically e-mailed search results to end users. Even though the amount of material identified and used in each search of the database was very small, the German court was persuaded that the systematic nature of the searches was sufficient to present repeated extraction and that the large number of searches constituted substantial use of the database, in the aggregate, even though each individual search yielded only very small amounts of information.

A single or limited instance of extraction or use of data collection content is not sufficient to permit recovery by the database maker under the directive (*C. Net* Case). However, a court in the Netherlands noted that in instances when there has been only limited extraction or reuse of content, the directive's terms could still apply if that small amount of content had substantial value to the user (*NVM v. De Telegraaf*). Thus, it appears that the decision as to whether there has been extraction or use of material sufficient to permit the content maker to obtain recovery under the directive is based on both the amount of content accessed and the value of that content to the party who accessed it.

The directive authorizes makers of data collections to enforce their rights of ownership for a period of 15 years from the date of its completion or from the date the database was first made publicly available, whichever date is later. There is, however, some uncertainty regarding what data constitutes the data of completion of a data collection. A court in the United Kingdom noted in passing, for example, that a dynamic online database that is continuously updated may never actually be completed. If that interpretation stands, the 15 year term under the directive would be continuously renewed for as long as the maker of the data collection continued to update the collection. The court in the United Kingdom was evaluating a dispute involving a Web site that presented horse-racing results that were updated with each race (*British Horseracing Board v. William H. Hill Organization*).

Internet content development and access generates diverse claims under the Database Directive. A court in Spain, for example, concluded that unauthorized reproduction and online posting of content from a database containing court case decisions and text of legislation was a violation of the rights established by the directive (Editorial Aranzadi). A French court determined that copying of subscriber information and online distribution of that information was a violation of database ownership rights (*France Telecom v. MA Editions*). In Germany, duplication of a Web site containing links to various other sites was deemed to be a violation of the directive (Kidnet/Babynet, Landericht Koln, August 25, 1999).

Some court decisions applying database ownership principles to Web content have generated confusing results. In Germany, a court determined that Web pages qualify as protected databases (*Baumarkt.de* Case). Another German court concluded that hypertext linking and framing were forms of reproduction and use of Web content; thus, parties who establish links or frames could be required to compensate the makers of the Web content they frame, or to which they link, under the database ownership directive (*Medizinisches Lexicon* Case). However, a different German court determined that music files made available online did not constitute databases and thus were not within the scope of the directive (*MIDI-Files* Case).

These somewhat conflicting interpretations of the directive in the context of Internet content raise some important policy issues that have yet to be resolved. For example, at least some of the Internet content that a few of the European courts have been willing to bring within the scope of the directive is also subject to protection under traditional copyright and trademark laws (e.g., Web pages). It does not seem appropriate that material protected by traditional intellectual property laws should also be covered by the database ownership rules. Such duplication of coverage appears to be both unnecessary and unduly restrictive on content use.

At present it is probably best to assume that a wide range of online content can qualify for protection under the directive. That scope of coverage has significant potential impact on Internet activities. Developers of Internet content can apply the rights granted by the directive as an additional tool to enable them to exercise control over their online material. Internet content users should remain mindful of this potentially broad scope and be cautious as they make use of online content.

The database ownership rights established by the directive must also be evaluated in the context of other legal initiatives under way in the European Community. For example, the EC has adopted a directive on electronic commerce (EC Directive 2000/31/EC of the European Parliament ["Directive on E-Commerce"]). In Recital 14 and Article 1 (5) of the directive on E-Commerce, there is a clear indication that it is the intention of the EC that the e-commerce transaction rules now being implemented are to be interpreted in ways that continue to give full effect to the requirements of the Database Directive. The EC thus recognizes that many of the electronic transactions that it intends to regulate under the directive on E-Commerce will involve data content that falls within the scope of the Database Directive, and it seems to establish the principle that the developing e-commerce transaction rules should be structured and interpreted in a manner consistent with the database ownership rights already created by the EC.

The rights established by the EC Database Directive have a significant potential impact on developers and users of digital content. Businesses operating in Europe and individual European citizens should pay close attention to the database ownership rules, as those rules are interpreted by the courts in the European countries. Recognize that the database ownership rules can be applied to data collections established by European companies and by non-European companies that are formally qualified to do business in a European country. The rules are also applicable to users of qualifying database content, even if those users are physically located outside of the European Community.

Data collection developers seeking to apply the database ownership rules should create records of evidence documenting the resources they applied to create and maintain their databases. The developers should also establish and maintain records that indicate the scope of all unauthorized use of their material. As we noted previously, in order to win a legal claim based on the Database Directive, a developer of a data collection must prove that it devoted significant resources to the establishment and maintenance of the collection and that the unauthorized use of the material was substantial.

Users of data subject to the Database Directive should also monitor carefully the continuing interpretation of the terms of the directive by the courts in the member countries. It is now clear that a wide range of material falls within the scope of the database laws. Collected information in print form, electronic databases, and Internet content are only some of the examples of the wide range of material reached by the directive. With such broad scope, a large number of information users, including publishers and Web surfers, can find themselves defendants in court cases brought under the directive. Accordingly, all users of data content developed or maintained by European parties should attempt to comply with all terms of use for that content established by the owner of the content. Users of data should also monitor their use to be certain that they know the sources of the diverse material that they access, that they can identify all uses that they make of the content, and that they are always aware of the amount of material that they are using (an element that is an important piece of evidence if there is ever a dispute over whether the use of the material was substantial enough to cause the terms of the directive to come into play).

8.2 Property law rights

The EC Database Directive established a set of ownership rights specifically directed toward the content of data collections. Some jurisdictions,
including various states in the United States, are considering a somewhat
different approach to manage digital content. Those jurisdictions are
beginning to apply well-established principles of property law to data and
to the computer files that contain the data. There is a growing trend
toward application of traditional property law ownership rights to digital
content. Intellectual property law rights (e.g., copyright, patent) were the
first property law principles applied to content on computer networks,
and they continue to be the primary source of rights for content developers. More recently, however, those content developers began to turn to
concepts of personal property law to supplement the rights they derived
from intellectual property law. Although the full scope and effectiveness of
property law theories in the digital environment has not yet been defined,
both developers and users of online content should at least be aware of the
arguments on both sides of this issue.

Under traditional property law, owners of various forms of property
were provided legal protection to control access to, and use of, that property by other parties. Traditional property law categorized property into
two groups: real property and personal property. Real property is generally defined as land (i.e., real estate) and the immovable fixtures associated
with the land (e.g., buildings). Personal property is basically defined as all
forms of property that are not real property or intellectual property.

The owner of personal property has the right to set the terms of access
to, and use of, the property. Access to, or use of, real property without the
owner's permission is commonly described as "trespass." Unauthorized
access to, or use of, personal property is commonly defined as "conversion" of the property. In many jurisdictions, the owner of property has the
right to use the legal system to force a party engaged in trespass or conversion to cease that conduct and to obtain monetary compensation (e.g.,
damages) for harm caused by the unauthorized use of the property.

It is now common for the owners of intangible property in electronic
form to attempt to assert both intellectual property law and personal
property law rights to help those owner manage access to, and use of, their
digital property. In the digital environment, the "property" at issue is both
the data located in computer files and the computer files themselves.
Thus, the developers of digital content stored on computers now routinely
take the position that the computer files they have created and the data

stored in those files are personal property owned by the developers. This claim of property ownership is generally presented in addition to the traditional copyright arguments that are based on intellectual property law.

Some courts have been called upon to begin to deal with these property law claims. In a dispute between eBay and Bidder's Edge, a U.S. federal court was asked to address a claim, raised by eBay, that information regarding product offerings (e.g., bid prices for products) available through the eBay online auction system were proprietary property of eBay and that Bidder's Edge could not access that property without the consent of eBay. Bidder's Edge made use of a search engine to obtain information regarding product prices at a variety of online auction sites, including that of eBay. Although the merits of this case were never fully resolved by the court, since the case was ultimately settled, the trial court did grant eBay a preliminary injunction barring access to the online information by Bidder's Edge. The court granted that injunction in part because it determined that eBay had a good chance of ultimately winning the case on the merits at trial. The court's decision to grant the injunction was appealed by Bidder's Edge; however, the appeal was withdrawn (as part of the settlement agreement between the parties) before it could be dealt with by the court of appeals.

A property law argument was also raised in a dispute between the House of Blues and a streaming media company, Streambox. In that case, House of Blues objected to the fact that Streambox established links to streaming media content available at the House of Blues Web site. Among other arguments, the House of Blues contended that the streaming media files were personal property of the House of Blues and that by linking to those files, Streambox had unlawfully converted the property without permission from the House of Blues. Before the court could act on the property law argument raised by the House of Blues, a settlement was reached in the case, so there is no court ruling to rely on from this case. The dispute offers, however, another example of traditional property law claims presented in the context of online digital content.

Another novel approach to property law in the context of Internet material is presented with regard to rights of ownership for Internet domain names. Traditionally, domain names have been managed under the terms of contract relationships (e.g., the contracts between domain name registrars and the domain name registries and the contracts between the registrars and the parties who register domain names), domain name management rules established by the Internet Corporation for Assigned

Names and Numbers, anticybersquatting laws, and trademark laws. More recently, some have raised the theory that rights of use for Internet domain names could be governed by personal property law principles.

A trial court in Virginia, for instance, ruled that rights of use for a domain name provided a property ownership interest against which a creditor could obtain a lien as partial satisfaction for a valid debt. That court decision was, however, ultimately reversed by the Virginia Supreme Court, which concluded that rights of use for domain names were governed by contract law rules, not personal property law (*Network Solutions, Inc. v. Umbro International, Inc.*). A court in New York applied a similar analysis, also concluding that an ownership interest in a domain name is a contract right, not a property right (*Zurakov v. Register.com*). A federal court in the United States has, however, suggested that it may yet be possible that domain names could, at least under some circumstances, be treated as personal property and subject to liens obtained by creditors (*Dorer v. Arrel*).

The extent to which property law principles of ownership will be applied to domain names remains highly uncertain. Owners of domain names should be aware, however, that the concept of domains as personal property will likely continue to be addressed by courts. As the early cases in this field suggest, application of property law principles to Internet domain names would have a mixed impact on domain name owners. As the cases suggest, if domain names are eligible for recognition as personal property, the owners of those domains would have additional legal claims to apply against parties who might misuse the domains. In such a setting, for example, an owner of a domain name could sue a cybersquatter for conversion of personal property, in addition to raising claims under anticybersquatting laws and under trademark law.

The cases also show, however, that if domain names are personal property, there can be some adverse consequences for the owners of those names. Creditors would be able to obtain liens against the use of those domain names as a means of satisfying debts owed by the owner of the domain. For example, if a company owes money to a supplier, the supplier could go to court and obtain a lien against the domain names of the company. Once in place, the lien would enable the supplier to obtain all or part of any revenue generated by the sale or license of the domain names. In addition, if a domain name is treated as personal property, it might have an impact on legal jurisdiction. If, for example, a U.S. company has no physical presence in Spain, but it chooses to register an Internet domain name in the country code top-level domain managed by Spain, that action alone might subject the U.S. company to the laws of Spain. The rationale behind such jurisdiction would be that the registration of the domain name in the Spanish TLD created legally recognized property in Spain. If such an argument were accepted, the act of registering the domain name in the foreign TLD could subject the U.S. company to the jurisdiction of Spanish law and the courts of Spain. It could mean, for example, that the domain name was now a form of property in Spain that could be subject to Spanish personal property tax, even if the domain name was not in active use.

Developers of digital content turn to property law theories to try to expand the scope of the legal protections available to them as they attempt to manage access to their material. Those developers use property law in cases when the traditional intellectual property law rules do not apply (e.g., for material that has economic value but does not qualify for copyright protection). Some forms of digital content, such as information, also carry important public benefits associated with access. It is likely that, over time, the law will recognize distinctions between conventional property and those forms of digital property, including information, that have widespread public value. Property law applied to digital property will continue to struggle to balance protection of ownership rights against public benefits associated with broader access to the property.

8.3 Property rights and competition law

Property rights, whether established by laws such as the Database Directive or by traditional legal principles of intellectual property or personal property, provide substantial rights of ownership, but those ownership rights are not limitless. Some of the most important limitations on the rights of property owners are established by the legal concepts of competition law and antitrust law. Competition law and antitrust law are generally designed to preserve the environment necessary for effective commercial competition. The primary public policy principle behind competition and antitrust laws is that protection of a competitive marketplace ultimately provides the best commercial environment for consumers and for continuing economic growth. Accordingly, most jurisdictions around the world prohibit property owners from asserting their ownership rights in ways that harm commercial competition.

When a property owner raises a legal claim alleging that another party misused some property, it is common to see the defendant in that case allege that the property owner has raised the legal claim in an effort to reduce competition. Thus, when we see litigation associated with digital property rights, we often see that the defendant in the case has raised a competition law or an antirust law claim against the property owner in response to the property claim. For example, in the dispute between eBay and Bidder's Edge discussed previously, Bidder's Edge responded to the eBay property law claim, in part, by arguing that eBay was engaging in unfair competition by trying to block access to the online auction information. Bidder's Edge sought to characterize the dispute as one involving unfair competitive practices by eBay, while eBay presented the case as a matter of property rights enforcement.

In some instances, claims of unfair competition can also be raised by the property owner. In those instances, the competition law claims are, in effect, supplements to the property law arguments. For instance, some courts in Europe have faced cases brought under the Database Directive that did not qualify for resolution based on that directive, but they have still provided relief for the plaintiff based on a finding of unfair competition. A French court, for example, concluded that a database containing calls for procurement contract bids did not qualify for protection under the directive, since the database was not the result of substantial investment of resources, yet the court ruled for the owner of the database based on a finding that the defendant's use of the database content was unfair competition (*Groupe Moniteur v. Observatoire des Marches Publics*). Database owners who raise claims under the Database Directive now commonly also raise unfair competition claims along with their claims under the directive, since courts have sometimes been willing to grant relief for unfair competition in some instances when they have been unwilling to apply the Database Directive.

Property ownership rights are intended to provide a legal basis upon which owners of property can receive fair compensation in exchange for rights of access and use associated with their property. Property rights are not intended to be the basis for impeding commercial competition. Owners of digital property should, thus, enforce their legal rights to obtain fair compensation for their property but should not enforce those rights as part of an effort to injure another business. Users of digital material should recognize that laws designed to protect competition can be valuable tools to ensure access to material that is essential for commercial competition. Those users should also be mindful, however, that if they misuse the digital property of other parties, those same competition laws can be applied against them, if their misuse of the digital property of another party constitutes a form of unfair competition.

To reduce the risk of antitrust or competition law liability, property owners should make sure that they can document a reasonable basis for limitations they impose on access to, or use of, their property. If, for example, a property owner can create a record of evidence that demonstrates that there were valid technical, safety, or economic reasons for restricting access to the property, the owner is less likely to face competition law claims based on the access or use restrictions, and that owner is more likely to prevail if a legal claim is raised. Conversely, a party who has been denied access to the digital property of another must build a record of evidence demonstrating that there are no reasonable factors supporting restrictions on access to the property and that the restrictions do, in fact, have an unreasonable, adverse impact on commercial competition.

8.4 Commercial transaction laws

Property rights theories also surface in the context of some of the commercial transaction laws that now address digital material. Laws applicable to the commercial sale and transfer of digital content have implications for property ownership rights. To the extent that commercial transaction laws either expand or reduce the scope of the rights of sellers of digital material, those commercial laws, in effect, alter the legal rights that owners of electronic content can exercise when they attempt to transfer that content. By altering those commercial rights, the laws have an impact on the ability of owners of digital property to control the terms of transfer of that property. In a sense, commercial transaction regulations thus influence the property rights of the sellers of the digital property. Two examples of this relationship between laws governing the rights and responsibilities associated with commercial sales are the EC Directive on E-Commerce and the Uniform Computer Information Transactions Act (UCITA).

The directive on E-Commerce was mentioned previously in this chapter. As was noted, that directive is designed to establish a framework of commercial transaction rules, for the EC, that will facilitate expansion of electronic commerce in the EC. A major portion of e-commerce involves sale of digital content. Rules that set the rights and responsibilities of buyers and sellers as to electronic transactions will have a major impact on access to, and distribution of, digital material (including information or data). As the countries in the EC enact their national legislation to implement the directive, they should recognize that those transaction rules must be consistent with the current property ownership and transfer rules, to the extent that the digital content transferred through the e-commerce transactions is subject to those property ownership rules. The EC appears to have recognized the need for consistency between its e-commerce regulations and its database ownership rules, since the E-Commerce Directive references the Database Directive. In the digital marketplace, regulations governing transactions and those defining property ownership rights are inextricably linked and must be effectively coordinated to avoid inconsistencies that would likely undermine the effectiveness of both sets of policies.

The United States also currently faces the issue of adjusting its commercial transaction laws to reflect the new realities associated with electronic commerce, while preserving a consistent framework of commercial property rights. Two states in the United States have adopted UCITA (Maryland and Virginia), and several other states are debating the merits of this statute. UCITA establishes a legal framework of rules for commercial transactions that involve sale of information products (i.e., computer software and all other materials that can be created, stored, or distributed using computers). In effect, UCITA establishes a set of legal rights applicable to commercial sale of digital material. UCITA remains highly controversial, in part because some users of digital material are concerned that it grants to the sellers of that material rights to control access and use of the digital material that extend beyond the traditional rights granted by copyright law. By altering the rights of commercial sellers of digital material, some opponents of UCITA contend that the statute effectively expands the control over the material that the sellers can exercise. Examined from a slightly different perspective, this criticism seems to make the case that UCITA, even though it is a commercial transaction law and not a property ownership law, effectively expands property ownership rights in the context of commercial sellers of digital content.

In an interesting trend, large-scale users of digital material (e.g., libraries, educational institutions) now often advocate greater reliance on traditional copyright law as an alternative to commercial transactions laws, such as UCITA. Although users of that type of content have often, in the past, been opposed to aggressive assertion of copyright laws, many of those users now believe that copyright law provides a more friendly environment for content users through its well-established principles such as fair use and the public domain. This shift reflects, at least in part, a recognition that legal principles providing rights of access to users are more clearly established under copyright law than they are under UCITA and the commercial transactions laws. In a sense, some of the large-scale users of copyrighted material who rely on the fair use exemption and public domain rights, and who were generally opposed to expanding copyright law by applying copyright rights to collections of data, now seem to suggest that expanded use of copyright law for data collections might have been a better policy approach, since that approach would have made it clear that the basic copyright principles that ensure some level of user access would apply to data collections.

Another interesting trend in advocacy caused by the debate over application of commercial transaction laws to digital content is the rise of consumer protection law principles in the context of electronic material. As property law and commercial transactions law focus more attention on digital material, users of that material direct greater attention to the fight for consumer protection regulations that apply directly to electronic transactions and to the distribution of digital content. For example, in the debate over adoption of UCITA, one common request by opponents of the statute is that it be amended to clarify that the law would not reduce the reach of already existing consumer protection regulations into sales involving digital material. These opponents are essentially arguing that if digital content has become widespread enough to be treated as a commodity product for the purpose of the sales laws governing its commercial distribution, then that content is widespread enough to justify application of basic consumer protection regulations (e.g., provisions governing warranties, refunds, exchanges) to the content as well.

8.5 Computer Fraud and Abuse Act and the Electronic Communications Privacy Act

Two federal statutes in the United States are now widely applied to support efforts to control access to digital property. The Computer Fraud and Abuse Act (CFAA) prohibits unauthorized access to material stored on certain computers. The Electronic Communications Privacy Act (ECPA) prohibits unauthorized access to electronic communications content. These statutes do not directly create property law rights; however, they have significant impact on assertion of control over digital content. In effect, they provide legal rights that are enforceable by computer system owners or by the owners of certain content against parties who access their computer content without permission. With this scope the CFAA and the ECPA provide additional legal rights to the developers and managers of qualifying digital material.

The CFAA bars unauthorized access to computers that are used by financial institutions or those that are used in support of a government function. The CFAA also prohibits unauthorized access to interstate computer networks, and it applies to unauthorized interstate access to any computer. This scope of coverage is broad. Accordingly, computer system operators have found room within the CFAA to apply the rights it grants to block parties who attempt to access and use the digital content stored and processed on their systems.

The ECPA prevents unauthorized disclosure of electronic communications message content and stored electronic communications messages. Authorization for disclosure of the message content must be obtained from the parties involved in the communications. The ECPA places obligations on public communications service providers to protect the privacy of communications content. The ECPA provides a legal vehicle that helps parties who engage in electronic communications to limit disclosure of the content of their communications. In this way, the ECPA offers legal rights that help parties who participate in electronic communications to control access to those communications.

8.6 Future trends

Expect increasing use of property law theories and commercial transaction law initiatives by developers of digital content in the future. Developers are likely to continue to see property rights as an important form of legal protection that can help them to manage more effectively their electronic material in the networked environment. As they continue to look for as many different legal theories as they can find to help them exercise control over their digital products, developers of those products will have every incentive to add basic property law arguments to their well-established intellectual property and contract law claims.

Expect users of digital material to make greater use of intellectual property law, consumer protection law, and competition law to protect their rights of access and use with respect to electronic material. As content developers expand their arsenal of legal weapons for use to manage their digital resources, users of those resources will find it necessary to expand their range of legal weaponry, as well, in order to maintain access to the material. For users, the most promising legal theories to support more open access to proprietary digital content will likely be the fair use and public domain principles of copyright law, along with expanded application of consumer protection remedies for digital products purchased by the public and competition law rights for businesses adversely affected by the assertion of proprietary rights by the owners of digital material.

In practice, both content developers and users should adopt strategies and procedures that position them to enforce their rights, while they also respect the rights of other parties involved in the creation and distribution of that content. For example, owners of digital material should effectively define their digital property and establish reasonable terms to govern access to, and use of, that property. Those owners should also make certain that their property descriptions and terms of use are clearly expressed in a readily accessible and understandable manner to all potential users of the material. Digital property owners should also make use of all reasonable technical means (e.g., access restrictions) available to them to prevent unacceptable use of their property. Reliance on technical self-defense measures may well provide a more efficient means of managing digital property than that offered by reliance on legal remedies enforced against unauthorized uses and users.

Digital content users should also develop and implement effective digital content strategies. Users should first and foremost respect the rights of content owners. This respect should be expressed by adoption of practices, for the use of digital material, that identify the rules of use established by the digital property owners and that enforce compliance with those rules. At the same time, those users should be sure to understand the legal rights available to content users and should be prepared to enforce those rights as appropriate.

It seems likely that courts, legislatures, and regulators will be required to balance the different legal strategies employed by content developers and users. Intellectual property and traditional property law theories are likely to be respected by legal institutions; however, the traditional limits imposed on those claims (e.g., fair use and public domain exemptions, as well as competition law requirements) will also be applied by arbiters of disputes between content developers and content users. Commercial transaction law principles will be applied more frequently to digital content, as the recognition that much of that content has reached commodity status becomes more widespread. As those commercial transaction rules develop, however, also expect increased use of consumer protection law and competition law principles to ensure that sellers of digital material are not able to use their legitimate rights of ownership in unreasonable ways. In this setting, both creators and users of digital material should understand, and be prepared to exercise, their full range of legal rights; yet both groups should also recognize that the public interest requires balancing of developer and user interests.

Selected bibliography

Readers interested in additional information regarding the EC Database Directive or material related to the property and commercial transaction law topics addressed in this chapter may want to consult the following resources:

Algemeen Dagblad v. Eureka, President District Court of Rotterdam, August 22, 2000.

Baumarkt.de, Oberlandesgericht Dusseldorf, June 29, 1999.

Berlin Online, Landgericht Berlin, October 8, 1997.

British Horseracing Board v. William Hill Organization, High Court of Justice, February 9, 2001.

C.Net, Kammergericht Berlin, June 9, 2000.

Computer Fraud and Abuse Act, 18 U.S.C. Sect. 1030 (1994).

Dictionnaire Permanent des Conventions Collective, Tribunal de grande instance de Lyon, December 28, 1998.

Dorer v. Arrel, 60 F. Supp. 558 (E.D. Va. 2000).

eBay v. Bidder's Edge, 100 F. Supp. 2d 1058 (N.D. Cal. 2000).

EC Directive 96/9/EC of the European Parliament and of the Council of March 11, 1996, on the legal protection of databases, Official Journal No. L77 of March 27, 1996.

EC Directive 2000/31/EC of the European Parliament of the Council of June 8, 2000, on certain legal aspects of information society services, in particular electronic commerce in the internal market. Information on the directive is available at http://europa.eu.int/ISPO/ecommerce/legal/documents/2000_31ec/2000_31ec_enpdf.

Editorial Aranzadi, Court of First Insistance Elda, July 2, 1999.

Electre v. T. I. Communication and Maxotex, Tribunal de commerce de Paris, May 7, 1999.

Electronic Communications Privacy Act, 18 U.S.C. Sects. 2510-2522, 2701-2711 (1996).

France Telecom v. MA Editions, Tribunal de commerce de Paris, June 18, 1999.

French Code de la Propriete Intellectuelle, Title IV, Book III.

German Information & Communication Services Act, August 1, 1997.

Groupe Moniteur v. Observatoire des Marches Publics, Cour d'Appel de Paris, June 18, 1999.

House of Blues v. Streambox (C.D. Cal., filed September 13, 2000), Complaint available at www.2001Law.com/article_385.pdf.

Kidnet/Babynet, Landgericht Koln, August 25, 1999.

KPN v. XSO, President District Court of the Hague, January 14, 2001.

Mars v. Teknowledge, High Court, Chancery Division (U.K.), June 11, 1999.

Medizinisches Lexicon, Landgericht Hamburg, July 12, 2000.

MIDI-Files, Landgericht Munchen I, March 30, 2000.

Netherlands Database Act, July 1999.

Network Solutions v. Umbro International, 529 S.E. 2d 80 (Va. 2000).

NOS v. De Telegraaf, Court of Appeals of the Hague, January 30, 2001.

NVM v. De Telegraaf, President District Court of the Hague, September 12, 2000.

Spanish Intellectual Property Act, amended 1998.

Suddeutsche Zeitung, Landgericht Koln, December 2, 1998.

Tele-Info-CD-Bundesgerichtshof, Federal Supreme Court, May 6, 1999.

Uniform Computer Information Transaction Act, information regarding the Act available at www.ucitaonline.com.

United Kingdom Copyright and Rights in Databases Regulations, 1997.

UNMS v. Belpharama Communication, Court of Brussels, March 16, 1999.

Zurakov v. Register.com, N.Y. Sup. Ct., N.Y. Cty., Case No. 600703/01, July 25, 2001.

9

Controls on technology development and transfer

A variety of legal restrictions are currently imposed on the development and distribution of information and telecommunications technology. Effective strategies for the development, sale, and use of that technology thus require an understanding of those restrictions and of the policy rationale behind the restrictions. This chapter focuses on two sets of constraints on technology development and transfer. One set of constraints includes the national and international regulations that restrict the export of technologies that have potential military applications. The second set of constraints consists of those constraints that apply to distribution of technology that can be used to infringe on copyright ownership rights.

These legal constraints limit the ability of technology developers to sell or otherwise distribute certain technologies (including computer software and hardware as well as telecommunications equipment). The constraints also restrict the ability of end users of the technologies to apply the technologies as they choose. By controlling both the development and the use of certain technologies, these regulations serve, in effect, as barriers to the development and the distribution of some forms of intellectual property and knowledge assets. The price paid by society for these legal constraints on specific intellectual assets is some level of reduced user access, delayed technological innovation, and reduced economic development.

The intended benefits to society from the technology restrictions are greater national security and additional protection for certain intellectual property owners.

Developers, distributors, and users of the intellectual property and knowledge assets targeted by these controls must understand the scope and form of the legal constraints directed toward their technologies. This chapter provides an overview of some of the key legal restraints on technology development and distribution. The chapter also discusses some of the methods that can be applied to manage technology within the parameters established by those legal restraints.

9.1 Technology export controls

Many of the most economically powerful countries in the world have laws that restrict the international distribution of certain technologies. Those legal restrictions can be organized into two categories: those that prohibit the distribution of certain technologies to specific countries and those that require prior approval of export of certain technologies based on the potential use of the technologies. Some countries have chosen essentially to prohibit export of technology to specific countries. For example, the United States applies such a policy to Cuba, and several countries apply similar trade sanctions to nations such as Iraq.

Restrictions on technology export also take the form of prior government approval for international distribution of technologies that have certain potential applications. One of the most common examples of this type of technology export control is the widely applied regulation of technologies that have military applications. Many different countries in the world require government approval prior to the international distribution of technology that can be used for military purposes. The prior government approval required for such exports is commonly described as an export license. The license is an authorization of the transfer by the government authorities. Licenses are required whether the regulated technology is exported for commercial or noncommercial purposes. The focus of these controls is on management of international access to controlled technology.

Certain technologies are primarily directed toward military use (e.g., weapons) and are thus relatively easy to regulate. Technology export licensing becomes more difficult to manage with regard to technologies that have both military and civilian applications. Such technologies are commonly described as "dual-use" technologies. In the past, governments attempted to separate military and civilian technologies, applying legal restrictions on development, distribution, and use of military technologies, but leaving civilian technologies relatively free from regulation. Advances in technology have made it difficult to separate civilian from military technologies. Many products developed purely for civilian commercial purposes contain components that could have military applications. For example, modern telecommunications switching equipment intended for purely civilian use commonly contains sophisticated computing components that could be used for military purposes. Space launch vehicles intended for use with communications satellites can also be applied for military launch applications. Computer game equipment meant for civilians now incorporates computing power sufficient to give it potential military uses as well. Virtually all technologies, particularly those in the computer and telecommunications fields, are dual-use technologies. As more products are recognized to incorporate dual-use technology, the regulatory regime applicable to technology exports is required to oversee a greater number of goods, exporters, and users. The greatest current challenge confronting export control authorities is the need to assess dual-use capabilities and to authorize international distribution of dual-use technologies in ways that promote economic growth yet simultaneously protect national security.

9.2 Overview of technology export laws

International distribution of technology with potential military applications is governed by a variety of international agreements and national laws. One of the key international agreements governing technology exports is the Wassenaar Arrangement (the Wassenaar Arrangement on Export Controls for Conventional Arms and Dual-Use Goods and Technologies). The Wassenaar Arrangement has been signed by 33 countries, but it does not establish specific export control regulations. Instead, the Arrangement creates policy goals to promote greater international security through the oversight of distribution of conventional weapons and dual-use technologies, and the Arrangement relies on the signatory nations to enact national laws that will support the policy goals. Essentially, the Wassenaar Arrangement establishes the international policy goals of promoting widespread international distribution of technology while simultaneously working to foster international security by managing the spread of munitions. The arrangement seeks to encourage global distribution of dual-use technologies, while permitting individual countries to try to manage the military applications of those technologies.

Countries that have signed the Wassenaar Arrangement committed to enact their own national laws consistent with the policy objectives of the arrangement. For example, Australia manages its technology exports through the Australian Defence and Strategic Goods List. The United Kingdom regulates technology exports under the terms of its Dual-Use Items (Export Control) Regulations 2000. Switzerland applies its Goods Control Decree. Israeli export controls are set through the Export Control Regulations. Technology export in Finland is controlled by the Finland Export Controls. In the United States, technology exports are controlled by the Export Administration Regulations, the Trading with the Enemy Act and the International Emergency Economic Powers Act, and the International Trafficking in Arms Regulations. Enforcement of U.S. export controls is a complicated task, since those controls are enforced by several different federal agencies, including the Department of State, the Department of Defense, the Department of Commerce, and the Treasury Department. This shared jurisdiction has made effective enforcement of the regulations difficult for both the U.S. government and for U.S. businesses. The different agencies have jurisdiction over the export of different items, and they are obligated to enforce different regulations. In addition, the various agencies involved in export control regulation have different areas of substantive expertise and different policy priorities. Collectively, these factors combine to make export control enforcement in the United States an inefficient and cumbersome process.

Export control rules generally carry with them substantial criminal penalties. Improper international distribution of technology with military applications can result in criminal penalties, including prison terms and fines, for all parties involved in the distribution process. Development and enforcement of effective compliance policies and procedures are thus essential for parties involved in international transfer or use of technology. Those who distribute any form of technology that can have military applications should be aware of the relevant export control requirements, even if the specific products to be exported are not intended for military uses. One example of the wide scope of U.S. export controls is the applicability of those controls to space launch vehicle and satellite technology. Companies in the United States have been fined for unlicensed export of space technology, even when the technology was exported for use to support the launch of telecommunications satellites and other forms of civilian use satellites.

Another example of the wide scope of export controls is the need for licensing for telecommunications switching equipment, technology that makes use of computer components that could have military applications, even when they are intended for civilian use. For practical purposes, it is best to assume that virtually all computer and telecommunications technologies are dual-use technologies, having both civilian and military applications. With that mindset, developers, distributors, and users of those technologies should understand export control regulations thoroughly and should be prepared to comply with the requirements of those regulations. Technology developers and users should be conservative as to export control compliance, erring on the side of assuming that their products fall within the scope of the rules instead of assuming that they do not.

9.3 Performance-based computer controls

Export of computers presents a prime example of the challenge of dual-use technology. Computers are increasingly powerful and useful, yet they continue to have both civilian and military applications. International distribution of computer equipment is widely regulated using a set of standards based on the performance capabilities of that equipment. This type of export control framework is often characterized as a performance-based regulatory regime. The most widely accepted performance standard is the computational processing speed of the equipment. Most of the leading economic countries evaluate computer performance, for export licensing purposes, based on the number of theoretical calculations per second that the computer is able to process. Computer authorities monitor how many million theoretical operations per second ("MTOPS") a computer can process. The greater the MTOPS number, the more useful the computer is deemed to be, and the more rigorous the rules applied to its export. The United States, for example, applies a standard of 85,000 MTOPS as the threshold at which an export license is required (1998 National Defense Authorization Act).

Performance-based export controls for computational devices are widely applied, yet their effectiveness is uncertain. The Wassenaar Arrangement applies performance-based standards to identify the computing equipment that requires export licensing. Export regulations in the

United States apply a similar regime under the terms of the National Defense Authorization Act of 1998. These standards become increasingly difficult to apply, however, as more powerful computing devices are commonly integrated into civilian devices (e.g., consumer electronics equipment, household appliances, toys). Continuing advances in the power of computing devices and their increasingly widespread integration into a wide range of products mean that more and more devices fall within the scope of export control restrictions applied to computers. In addition, rapid advances in computer technology force frequent modifications to the performance-based standards in order to accommodate the technological advances.

Another difficult challenge for performance-based export controls is the increasing use of networks of computers. In a networked setting, a set of less powerful computers can be connected to create far greater computational capability. Use of networking capability can circumvent computing performance standards by enabling parties to acquire multiple devices that are not restricted by export controls and interconnect those less powerful devices to create far more powerful computing systems. In a setting where use of computer networks is common, performance standards applied to individual computers is a far less effective means of controlling computer use.

An increasing number of computer industry observers now question the effectiveness of performance-based export controls for computing devices. For example, a report commissioned by the Center for Strategic and International Studies concluded, in part, that performance-based export controls for computing equipment were no longer effective and that alternative strategies for export regulation should be explored ("Computer Exports and National Security in a Global Era," a report of the CSIS Commission on Technology Security in the Twenty-First Century).

Although some observers question the effectiveness of performance-based standards for computer export controls, those standards continue to be in place. Parties involved in a wide range of industries must be aware of those standards and must establish effective policies and practices for compliance. In today's marketplace, it is often easy to forget just how many different devices contain powerful computing capabilities. As computing becomes ubiquitous and less visible, a wider range of products and product developers, distributors, and users will be brought within the scope of export controls. Manufacturers, distributors, and users of devices that have computing capability should focus on the capabilities of their devices, not their intended purposes. Export controls apply to devices based on their performance capabilities; thus, if a product has powerful enough computing functions, it is regulated, even if it is intended for non-military applications. Everyone involved with goods that make use of computing capability should be aware of technology export regulations and should monitor the continuing changes in those regulations. In addition, all those affected by computer performance standards should play an active role in the review and modification of those standards to help ensure that the standards strike an appropriate balance between security and economic development.

9.4 Encryption controls

A subject of particular attention, from a technology export perspective, is international distribution of cryptographic systems used for data encryption. Encryption technology provides one of the clearest examples of the challenges associated with management of dual-use technologies. Demand for encryption systems, using both hardware and software solutions, increased dramatically with the rise of electronic commerce. Yet law enforcement and national security authorities fear that widespread use of powerful encryption poses a threat to security interests. Since cryptographic systems and technologies have clear potential military applications, virtually all export control regulations apply to international distribution of encryption capability, whether in the form of hardware or software.

The Wassenaar Arrangement applies to the export of encryption technology. Signatories to the arrangement permit the export of cryptographic technology, subject to licensing restrictions that vary from country to country. Some countries adopted very liberal export rules for encryption technology, allowing widespread international distribution with relatively few licensing requirements. Finland, for example, permits essentially unlimited international distribution of cryptographic technology provided that it is incorporated into mass market hardware or software products (Export Control of Dual-Use Goods Act of Finland (562-96) and National Cryptography Policy Guidelines [October 1998]), and Germany applies a similar approach (General License Number 16). Ireland permits export of mass market encryption software (up to 64-bit key software) without a license (Ireland's Policy on Cryptography and Electronic Signatures).

Australia provides a general exemption from licensing requirements for encryption software that is part of the public domain (e.g., Open Source software) (Australia Defence and Strategic Goods List). The United Kingdom and Sweden provide for exemptions from the licensing requirement when the exported encryption technology is to be used for personal, noncommercial purposes (U.K. Dual-Use Items [Export Control] Regulations 2000 [SI 2000/2620] and Swedish Export Regulations, SFS 1998).

Other countries adopted more restrictive licensing regimes for export of encryption systems. For instance, Israel requires a case-by-case review of all encryption exports to determine whether a license is required (Israeli Encryption Control Regulations). Japan generally requires a license for export of encryption systems valued at more than 50,000 yen, and countries such as Myanamar and Belarus require licenses prior to export of virtually all cryptographic products (Computer Science Development Law of Myanamar and Resolution No. 218 [1997] of the Council of Ministers of the Republic of Belarus).

In some countries, export control regulations applied to encryption technology have evolved dramatically. France is an example of such an evolution. French laws governing export and use of cryptographic systems were initially restrictive, but over time they have been modified to be significantly more liberal (Export Control Law of July 26, 1996, and Implementing Decrees of February 28, 1998, and March 17, 1999). Additional liberalization of the French laws applied to distribution and use of encryption is now under consideration (Bill No. 3143 on the Information Society).

China is another example of a country with rapidly evolving policies toward encryption technology. Chinese regulations continue to be highly restrictive with respect to the distribution and use of cryptographic products in China. The government has, however, taken steps to reduce the restrictions on international export of cryptographic systems from China to the rest of the world. For example, China now permits the international export of encryption technology, without a license, provided that the encryption function is only an ancillary aspect, not a core function, of the exported product (State Council Order No. 273).

Application of export controls to encryption technology has been particularly challenging for two reasons. First, the technologies involved include both computer hardware and software; thus, export controls applied to encryption affect developers, distributors, and users of both sets of computer-related products. Second, encryption software is easily distributed to a wide range of users around the world; thus, the transactions

are difficult to monitor, and they involve a diverse groups of parties, ranging from large businesses to individual people. Electronic transfer of
encryption software is common and encryption capability is widely integrated into various mass market software products (e.g., Internet brows-
ers). These characteristics of the development, distribution, and use of
encryption systems made those systems extremely difficult to manage
within the traditional export control framework.

Expect export controls applied to encryption technologies to continue
to evolve rapidly for the foreseeable future. The conflicting policy objectives associated with use of encryption make it a technology that will
continue to be subject to diverse regulatory treatment. Developers, distributors, and users of encryption products must continue to comply with
the established framework of export controls. Recognize, however, that
the framework of controls will change over time, and all compliance procedures must be structured so that they can be modified quickly to meet
altered regulations.

9.5 Knowledge controls

In addition to regulating the export of technology products, export controls also limit the international distribution of knowledge and information pertaining to technology that has potential military applications. Just
as the international export of certain tangible goods requires prior government approval, so too does the international distribution of certain information or knowledge. These restrictions apply to distribution of the
regulated knowledge using any medium, including written and verbal *or a*
communications.

To illustrate the scope of these knowledge controls, consider the field
of cryptography. Assume that Professor X has been conducting research
into computer cryptography systems. He writes an article based on his
research and would like to publish that article. Professor X must first recognize, however, that he may be required to obtain a government license
prior to publishing the article. The publication of the article could constitute international distribution of the information, and if the substance of
his research has potential military applications, then Professor X will be
deemed to be exporting regulated information. Similarly, Professor X may
be required to obtain an export license if he plans to teach a course cover-
ing the restricted material. A license could be required in that instance if

the course material will be accessible to foreign nationals (e.g., as students or through international distribution of course materials).

Several court cases in the United States have addressed the issue of the government's ability to restrict international distribution of knowledge on subjects covered by the U.S. export control regulations. For example, in the court case *Bernstein v. U.S.*, federal courts reviewed the constitutionality of the U.S. government's interpretation that the export control regulations required a license before a graduate student could post his research into certain cryptographic systems online. In the *Bernstein* case, the courts granted a preliminary injunction temporarily preventing enforcement of the regulations against Bernstein. The courts ultimately concluded that changes in the U.S. export control rules for encryption technology implemented while the *Bernstein* case was pending made Bernstein's publication of his research legal without a prior export license, so the courts concluded that they were not required to resolve fully, at this time, the legal issues raised by Bernstein.

Another federal court in the United States addressed the issue of export controls applied to academic publication of cryptographic information. In *Junger v. Daley*, a university professor argued that export controls placed an unconstitutional burden on his ability to teach cryptographic information to classes containing foreign students and on his ability to publish his course materials. On appeal in the *Junger* case, the federal court concluded, as did the trial court in the *Bernstein* case, that distribution of information on cryptography and encryption, in an academic setting, had greater constitutional protection than commercial distribution of the same material. Courts in the United States have thus been willing to permit more liberal academic distribution of information governed by U.S. export control regulations than commercial distribution.

Export control laws regulate the international flow of certain information and expertise. This gives those laws very broad potential scope. Businesses and individuals involved in the development, use, or study of technical information that has potential military application should be aware of their exposure to potential liability under export control regulations. Compliance with those regulations is, in some ways, a greater challenge with regard to the flow of knowledge than it is in the context of tangible goods. Regulated information and knowledge can be distributed in many different forms (e.g., print, electronic, verbal), and all forms of distribution are subject to regulation. The rapid global dispersal of information facilitated by the Internet and other forms of communications makes it very difficult to manage effectively export control compliance with regard to knowledge and information. All controls and procedures implemented to manage export control compliance for goods and products should also be consistently applied to knowledge and information. If there is reason to believe that regulated information has been improperly distributed without an export license, the disclosing organization should promptly report the disclosure to the government authorities who enforce export control laws. Prompt disclosure and cooperation with government authorities can significantly reduce any penalties that may be associated with the violation.

9.6 Restrictions on pass-through exports

Export control rules generally prohibit distribution of regulated products to parties who, in turn, export the products without appropriate authorization. For example, a company that makes a lawful domestic sale of a product to a party who, in turn, makes an unlawful international distribution of the product can be held accountable for the unlawful international export (along with the party who actually exported the product) if the original seller knew, or had reason to believe, that an unlawful export of the product would take place. This aspect of export control rules places a burden on parties who distribute regulated products, even if they do not engage in direct international distribution of the products. These distributors have a legal responsibility to take reasonable measures to ensure that their products are not redistributed without appropriate export authorizations.

The duty to manage redistribution of products governed by export controls is made more challenging by the fact that products that contain regulated components can be disassembled or reverse-engineered and the regulated components redistributed. An exporter of a product that contains a regulated component can be held accountable for export violations if that regulated component is redistributed without proper government approvals. For example, if a software developer distributes a product that contains an encryption capability (even if that capability is not the primary function of the software), and if that encryption capability falls within the scope of export control rules, the developer will generally be accountable if the buyer of the software violates export controls when it redistributes the software or the encryption component. The developer may be held accountable even if it did not approve the redistribution or, in some circumstances, even if it was not aware of the redistribution. The standard applied to the developer in that situation is essentially whether, given the particular facts of the arrangement, the developer actually knew of the redistribution or if the developer should have anticipated the redistribution, given the facts of the situation.

Prohibitions against redistribution of controlled material apply to knowledge and information in addition to tangible goods. If, for example, a consulting firm writes a report regarding computer equipment that falls within the scope of export control regulations, the firm can be held liable for improper international redistribution of material contained in that report by the client, even if the consulting firm did not authorize the redistribution and was not aware of the redistribution. If the consulting firm did not take adequate care to prevent the redistribution, it can be held liable for the violation, along with the client who actually distributed the controlled material.

Developers and users of goods and information governed by export control regulations should apply a basic strategy to ensure effective compliance with the obligation to minimize the risk of unlicensed pass-through of regulated material. All contracts executing the transfer of products or information that have potential military applications should clearly state that the products or information will not be redistributed without receipt of all required government approvals. Parties who transfer such material should create a record demonstrating the precautions they have taken to ensure that the material will not be redistributed improperly. If the party who transfers the material has reason to believe that unlawful redistribution of the material has taken place, that party should inform the government authorities responsible for enforcement of export control regulations.

9.7 Digital Millennium Copyright Act and anticircumvention

In the United States, copyright law was modified by the enactment of the Digital Millennium Copyright Act (DMCA). One particular provision of the DMCA is relevant to this discussion of legal constraints on technology development. The DMCA includes a prohibition on the development, distribution, or use of technologies that can be applied to defeat or override copyright protection systems. This provision of the DMCA is often referred to as the "anticircumvention provision." The anticircumvention provision makes it illegal to create, share, or operate a technology that can be used to defeat measures applied to protect copyrighted material, even if there has been no actual copyright infringement and even if the technology involved has noninfringing, lawful uses.

The anticircumvention prohibition represents an important shift in copyright law policy. In the past, copyright law focused on providing remedies for harm caused to copyright owners by misuse of their property. Traditional copyright law provided a legal mechanism to force infringers to cease their misappropriation of the copyrighted material and to permit monetary compensation to the copyright owner for damages caused by the infringement. With the anticircumvention provision, the DMCA gives U.S. copyright law the ability to bar the development and use of certain technologies, even when there has been no proof showing that the technology in question has caused any actual harm to any copyright owner.

The DMCA's anticircumvention provision is highly controversial. Several courts in the United States have been asked to examine the constitutionality of the provision. Results of that judicial review have been mixed. Some federal courts have concluded that the anticircumvention provision is not an unlawful interference with the constitutional right of free expression (*Universal Studios v. Reimerdes* and *Universal Studios v. Corley*). A federal court in the United States refused to issue a declaratory ruling to permit a university professor to publish information describing a system that can be used to circumvent a new copyright protection software for music (*Felton v. Recording Industry Association of America*); however, the parties did reach a negotiated settlement. A state court in California, when reviewing a trial court decision regarding online posting of the DeCSS DVD decryption software, concluded that interpreting the DMCA to prohibit online posting of the software in question was unconstitutional (*DVD Copy Control Association v. Bunner*).

The anticircumvention provision of the DMCA is also being applied by the U.S. government against foreign software developers and distributors. The Russian software developer, Dimitry Sklyarov, was prosecuted by U.S. authorities under the DMCA based on his online posting of software that can be used to read and duplicate certain electronic book content (*U.S. v. Sklyarov*). Although the software development and posting took place outside of the United States, and although it does not appear that

Mr. Sklyarov actually infringed on any copyrighted material, U.S. authorities arrested Sklyarov when he came to the United states and moved to prosecute him for violation of the DMCA's anticircumvention provision.

The *Sklyarov* case should be a warning for software and hardware developers and distributors in all parts of the world. Although they may be physically located outside of the United States, they could still face criminal prosecution under the DMCA to the extent that the software or equipment that they develop and distribute falls within the scope of the anticircumvention provision.

The anticircumvention provision has also been used against alternative systems of online content distribution. For example, copyright owners have used the provision against peer-to-peer content-sharing systems (e.g., Napster). Copyright owners argued that although peer-to-peer systems have many lawful, legitimate uses, they can also be used to facilitate the unauthorized distribution of copyrighted material (e.g., music recordings, software, video content). To the extent that those content-sharing systems can support unauthorized distribution of the copyrighted material, many copyright owners contend that the systems violate the anticircumvention provision of the DMCA. This argument has not yet been fully addressed by courts, but it has been raised in several U.S. copyright law cases (*RIAA v. Napster, Zomba Recording v. John Deep, Twentieth Century Fox Film Corp. v. Scour, Inc., MGM Studios, Inc. v. Grokster Ltd.*).

The DMCA was primarily intended to be a statute that would help to clarify the copyright protection to be afforded to digital material. The anticircumvention provision of the act seems, however, to expand copyright protection in the United States to the point where it can impede technological innovation and rights of free expression. At present, the DMCA is thus both a tool to help owners of intellectual property to protect their property from unauthorized use and a weapon that is being used against the developers of other forms of intellectual property (e.g., decryption systems, content-sharing systems). The future challenge is to continue to protect effectively intellectual property ownership rights without impeding technological innovation and open access to digital content.

9.8 EC Cybercrime Treaty

The European Community is in the process of ratifying a Cybercrime Treaty. The treaty begins the process of clarifying the scope of online criminal conduct. One of the steps that the treaty takes is to make copyright infringement a criminal offense. In addition, the treaty makes aiding or abetting copyright infringement a criminal offense. Thus, both parties who engage in copyright infringement and those who contribute to the process of infringement can be subject to criminal penalties under the terms of the treaty, in addition to traditional civil penalties.

In that environment, parties who manufacture or distribute technologies that facilitate copyright abuse can be subject to criminal penalties. This scope of coverage makes the Cybercrime Treaty similar to the DMCA of the United States, since both laws expand the range of penalties that can be applied to those who create or distribute technology that can be used for copyright infringement, even when that party did not directly engage in copyright misuse.

9.9 Technology taxes and fees

As an alternative to legal restrictions on the distribution or use of technologies, some jurisdictions apply economic charges (e.g., taxes or fees) to the manufacturers or distributors of the technology. For example, Germany assesses fees against equipment that can be used to duplicate copyrighted material (e.g., VCRs, cassettes, CD burners). The revenue derived from those fees is used to compensate copyright owners. Under the German system, technologies that can be used to create pirated copies of copyrighted material are not restricted; however, the manufacturers of those technologies are held accountable to compensate the copyright owners. The equipment manufacturers, in turn, commonly pass those costs on to the consumers who buy the equipment. In this way, all purchasers of technology that can be used to duplicate copyrighted material bear part of the cost of compensating copyright owners for unauthorized use of their content.

This approach makes use of an economic structure to supplement legal restrictions. It attempts to find ways to compensate intellectual property owners for misuse of their property instead of relying on legal constraints on technology. This approach strives to spread the cost of misuse among a range of product users instead of denying them access to the product. Under this strategy, consumers continue to have access to the products they demand; however, they are required to compensate content providers for losses resulting from piracy. Expect additional initiatives of this sort in the future as more jurisdictions search for alternative means of protecting content while encouraging continuing technological innovation.

9.10 Technology controls: Trends and strategies

Governments will continue to be encouraged to attempt to restrict and to control different technologies. It is often tempting for governments to believe that they can effectively accomplish various policy objectives (e.g., enhance national security, assist law enforcement efforts, protect intellectual property) by prohibiting or limiting the use of some technologies. Technology restrictions are often viewed by governments as a direct method of preventing conduct enabled by the technology. This perception is at the heart of technology controls. It seems likely, however, that in the future, governments will learn, from direct experience, that technology controls imposed by law are not an efficient or particularly effective means of public policy implementation.

The efforts to prohibit development, distribution, or use of specific technologies are likely to be unsuccessful in the long run. To the extent that more technologies are subject to government control, management of those controls will likely be more difficult as the number of products at issue becomes greater. If the number of products affected by technology regulation is large, more parties will be aware of the regulations and will have an economic interest in limiting the scope of those regulations. As regulated technologies become integrated into mass market products, it is increasingly difficult to enforce restrictions on their distribution and use. In addition, as more countries are able to support their own technology industries, sources of regulated technologies proliferate and it becomes more difficult to limit the distribution of those technologies.

Developers and users of restricted technologies should attempt to comply with the legal constraints applied to those technologies. Export controls and other forms of technology regulation carry the force of law and must thus be respected. Those laws continue to evolve, however, and compliance is made more difficult as the scope and form of the technology restrictions rapidly change.

In the future, we are likely to see greater reliance on use of economic incentives instead of classic legal prohibitions. Economic incentives to alter the form and use of some technologies are likely to prove to be more effective in the long term than laws or regulations applied directly to the development or use of the technologies. For example, the tax assessed against equipment manufacturers by Germany may prove to be a more effective and more manageable regulatory framework for copyright protection than the anticircumvention prohibition of the DMCA in the United States.

As technology advances, governments may begin to see that regulations aimed at blocking the development of specific technologies will be increasingly inefficient and ineffective. With many different sources of technology around the world, it is now more difficult to regulate access to technology than it was in the past, and that trend is likely to continue in the future. The future will most likely show that economic forces and incentives stand a better chance of influencing the use of technologies than do restrictions imposed by law. Owners and users of intellectual property should abide by the current framework of legal controls on certain technologies, but they should also prepare for a future when those legal constraints will be less prevalent and economic incentives will be more widely applied. Those who assume that laws alone can control the development and use of new technology are likely to be surprised in the future. Enterprises that construct their business plans based on the assumption that legal barriers to new technologies will prevail in the future are likely to find themselves at a distinct competitive disadvantage in the future.

Selected bibliography

Australian Defence and Strategic Goods List, www.dao.defence.gov.au/exportcontrols/dld_dsgl.html.

Australian Export Controls, www.dmo.defence.gov.au/DMU/function.cfm?function_id=50.

Belarus Resolution of Council of Ministers, No. 281, March 18, 1997, http://projects.sipri.se/expcon/natexpcon/Belarus/belcust.htm.

Bernstein v. U.S., 947 F. Supp. 1288 (N.D. Calif. 1997), at www.eff.org/pub/Legal/Cases/Bernstein_v_DoJ and Ninth Circuit Case No. 97-16686, May 6, 1999, September 30, 1999, and April 12, 2000 Orders.

Canadian Export Controls on Cryptographic Goods, www.dfait-maeci.gc.ca/~eicb/notices/sec113-e.htm.

China State Council Order No. 273, Commercial Use of Password Management Regulations, October 15, 1999.

Computer Exports and National Security in a Global Era: A Report of the Center for Strategic and International Studies Commission on Technology Security in the Twenty-First Century, May 2001 (Executive Summary at www.csis.org/tech/computer_execusum.pdf).

Computer Science Department Law of Myanamar, SLORC Law No. 10/96, September 20, 1996, as amended by Law No. 3/98 of February 23, 1998.

Danish Encryption Policy, April 7, 2000, www.fsk.dk.

DVD Copy Control Association v. Bunner, Case No. CV 786804, Santa Clara Superior Court, California, filed 1999.

European Community Convention on Cybercrime, ETS No. 185, open for signatures November 23, 2001, http://conventions.coe.int.

European Council Regulation No. 1334/2000, setting up a community regime for the control of exports of dual-use items and technology, Official Journal L159, 30.1.2000 (June 22, 2000).

Felton v. RIAA (D. N.J., filed June 6, 2001), www.eff.org/Legal/Cases/Felton_RIAA/20010606_eff_complaint.html.

Finland Export Control of Dual-Use Goods Act, 562/96, www.vn.fi/ktm/eng/2/Exportcontrols/1.htm.

French Bill on the Information Society, No. 3143, www.internet.gouv.fr/francais/textesref/pagsi2/lsi.htm.

French Law on Encryption of July 26, 1996, and Implementing Decrees of February 28, 1998, and March 17, 1999, www.telecom.gouv.fr/english/activ/telecom/nloi17.htm.

German License Number 16, August 31, 1999, www.sicherheit-im-internet.de/download/ag_16.pdf.

Ireland Policy on Cryptography and Electronic Signatures, June 1998, www.irlgov.ie/tec/communications/signat.htm.

Israeli Encryption Control Regulations, www.mod.gov.il/modhl/encryption/enc_b.htm.

Junger v. Daley, 209 F. 3d 481 (6th Cir. 2000).

MGM Studios, Inc. v. Grokster Ltd. (W.D. Cal. filed October 2, 2001).

RealNetworks v. Streambox (W.D. Wash. Filed January 18, 2000).

RIAA v. Napster, Case Nos. 00-16401, 00-16403 (9th Cir. 2001).

Swiss Goods Control Decree, June 25, 1997, www.admin.ch/bawi/f/knotroll/gkve.pdf.

Twentieth Century Fox Film Corp. v. Scour (S.D. N.Y. filed July 20, 2000).

U.K. Department of Trade and Industry 19989 White Paper on Strategic Export Controls, www.dti.gov.uk/export.control.stratex.

U.K. Dual-Use Items (Export Control) Regulations 2000, SI 2000/2620, www.hmso.gov.uk/si/si2000/20002620.htm.

Universal Studios v. Corley, Case No. 00-9185 (2nd Circuit 2000), www.eff.org/IP/Video/MPAA_DVD_cases/20010501_ny_hearing_transcript.html.

Universal Studios v. Reimerdes, Case No. 00 Civ. 0277 (LAK) (S.D.N.Y. August 30, 2000), www.eff.org/IP/Video/MPAA_DVD_cases/20000830_ny_amended_opinion.pdf and www.eff.org/IP/MPAA_DVD_cases/20000817_ny_opinion.pdf.

U.S. Digital Millennium Copyright Act, 17 U.S.C. 1202. DMCA overview at www.loc.gov/copyright/legislation/dmca.pdf.

U.S. Export Administration Regulations, 15 C.F.R. 768 et seq. (1996).

U.S. International Trafficking in Arms Regulations, 22 C.F.R. 121 (1996).

U.S. National Defense Authorization Act (1998).

U.S. Trading with the Enemy Act and International Emergency Economic Powers Act, 50 U.S.C. 1701 et seq. (1994).

U.S. v. Sklyarov, Case No. 5-01-257 (N.D. Cal. filed July 10, 2001), Complaint available at www.eff.org/Legal/Cases/Felton_v_RIAA/20010606_eff_complaint.html.

Wassenaar Arrangement on Export Controls for Conventional Arms and Dual-Use Goods and Technologies, www.wassenaar.org/docs/index1.html.

Zomba Recording Corp. v. John Deep, Case No. 01 CV 4452 (S.D. N.Y. filed May 24, 2001).

10

Valuing and managing
knowledge assets

The economic value of intellectual property and other forms of intellectual assets is now widely recognized. Developers and users of those assets now treat that property in much the same manner that owners of traditional property (e.g., land, personal tangible property) have long applied to their property. In addition, the rise of the Internet and the expanded use of computer networking, in general, create an environment that both enhances and diminishes the economic value of intellectual assets. This chapter discusses some of the issues and principles applied to the valuation and management of intellectual assets.

10.1 Valuation

A fundamental challenge associated with management of intellectual assets is the issue of identifying appropriate estimates of their economic value. Accurate assessment of the value of an intellectual asset is a necessary first step toward effective decision making as to investment in development and commercialization of the asset. Several different models are commonly applied to estimate the value of intellectual assets. Each model has its strengths and its weaknesses. The appropriate choice of a valuation model depends on the specific facts associated with the situation. The type of intellectual asset, the history of its development, and the objectives of the developer of the asset are critical elements affecting the selection of a valuation model.

One common valuation model is the cost-based approach. This model seeks to estimate the costs incurred in the development of the asset, plus some additional amount for profit. The theory behind this valuation approach is that the value of an asset is the amount of the investment in its creation, plus some amount of gain as compensation for the application of the resources. Cost-based valuation models tend to be backward-looking, historical valuation models. They focus on recovery of the investment associated with the development of the asset. Some cost-based valuation models focus on the cost to replace the intellectual asset instead of the cost of development for the asset.

One of the key advantages of the cost-based valuation model is the fact that it is generally readily quantifiable. Historical costs incurred in the development of a new intellectual asset can be easily documented. The precise nature of the cost analysis makes it relatively easy to document a cost-based value for virtually any intellectual asset. In sum, the cost-based model for asset valuation is often selected largely because it is easy to apply and can generate asset values that can be documented without relying upon projections or estimates of future economic conditions.

The historical focus of the cost-based valuation model also provides the core of its weakness as a model, however. The emphasis on costs already incurred often ignores, or understates, the potential future benefits that can be derived from the asset. A cost-based analysis often ignores future cost reductions that can result from use of the asset by the developer. The cost analysis also often ignores potential earnings that can be derived from the asset. Licensing revenue, for example, is commonly left out of the cost valuation model.

Some versions of cost-based value analysis also include estimates of costs saved by the development of the asset. If, for example, a manufacturing company develops a proprietary technology that helps it to reduce the expenses it incurs in its manufacturing process, the costs saved by the manufacturer as a result of the fact that it possesses the technology can be applied to reduce the total costs associated with the creation of the technology. These cost savings can thus be used to offset some of the development costs. The rationale for this approach is that cost savings associated with an intellectual asset are both quantifiable and relevant to value estimates for the asset. However, these models generally do not project the costs savings into the future, focusing instead on the savings derived from the technology in the past.

Another popular valuation model is market-based analysis. This approach attempts to estimate a value for the asset based on the market value of comparable assets. Market-based asset valuation models tend to focus on current price data for comparable assets now commercially available. In order to use a market-based valuation model effectively, there must be comparable assets already in commercial distribution. Market prices can only provide useful estimates to the extent that they are based on current prices and are associated with assets that are substantially similar to the ones for which the estimate is being developed. Innovative technologies that are not similar to any other technology currently in the marketplace can generally not be valued using a market analysis, since no accurate market data are available. It is, however, this type of truly innovative technology that has the greatest long-term potential value.

The market-based valuation model identifies the market value already established for a particular intellectual asset. The advantages of this model are clear. It makes use of reliable valuation data, since it captures actual price information. Use of actual price information provides the best possible data to estimate value. Based on this type of actual data, the resulting valuations tend to be relatively easy to document, and they provide a greater degree of comfort as to accuracy.

The market valuation model also has significant disadvantages. The primary disadvantage is that it does not work well in situations where there is no identical or similar technology already in the marketplace. Absent a comparable asset already in commercial circulation, there is no applicable current price information. Without that current market pricing data, the market valuation model is not effective. In addition, the model does not work well for assets subject to rapid price shifts. If, for example, a technology becomes outdated quickly, the market value for that technology will fall rapidly. Under those circumstances, a current market valuation carries little value as an estimate for future value.

Some parties apply an income-based valuation method. This system establishes a value for intellectual assets based on estimates of the future income they can be expected to provide to the developer. Some parties enhance this model by including an assessment of royalty payments that the developer can avoid paying because it owns the asset. Income-based models are forward looking, basing their valuations on predictions of future gains. They attempt to establish an estimate of the discounted cash flow value of revenues that can be derived in the future from the intellectual asset.

More recently developed asset valuation models attempt to reflect some of the decisions and choices that owners of the asset make as they develop and commercialize their intellectual assets. For example, some observers attempt to value the "options" (i.e., the choices) available to the asset owner as it commercializes the property. Option analysis is commonly applied to value the right associated with the opportunity to sell an asset at a future time, when that asset has varying market prices. An options approach to valuation is thus appropriate for intellectual assets, since they tend to be subject to significant variations as to future market price. The estimated value of those options is then incorporated into the assessment of the overall value of the underlying intellectual asset. Another approach is to account for the various different choices that the owner can make as to use of the intellectual asset (e.g., obtain foreign patents, let patents lapse) and to estimate the value of each of those choices. The values of those possible uses are then integrated into the overall value of the asset itself. Each of these valuation models attempts to provide some estimate of the value associated with the different opportunities associated with the use of intellectual assets that are available to the owners of those assets.

When assessing the value of an intellectual asset, it is important to take into account the potential value that an asset can carry as a result of the fact that it is under your control and is thus not available for use by a competitor. Under some circumstances, for example, the expense of obtaining and maintaining a patent may be worth it for an organization if the patent protection keeps a competitor from making use of the innovation. This value could exist even if the patented technology was not essential for use by the patent owner. Intellectual asset valuation is thus affected by both the value of the asset when used by its owner and the value, to the owner, of preventing competitors from obtaining access to the asset.

Perhaps the most effective valuation strategy is use of multiple valuation models. By combining more than one of the standard valuation models, developers of intellectual assets have a better chance of deriving accurate value estimates. Use of multiple valuation models does, however, make the process of value estimation more complex and more difficult to perform accurately. Even when multiple asset valuation models are applied, there continues to be substantial margin for error as to the specific values that are derived. The intellectual assets that have the greatest economic value are those that can be most effectively protected. For example, technology protected by patents is commonly more highly valued than other innovations that cannot be patented. Intellectual assets that are subject to less effective legal protection (e.g., trade secrets) may have substantial economic value, but that value is often discounted, since the owners are not able to exercise full control over the assets.

10.2 Licensing

After evaluating the commercial value of an intellectual asset, the owner of the asset must make decisions regarding grant of rights of use and potential transfer of ownership. To transfer ownership or rights of use associated with intellectual assets, contracts known as "licenses" are established. Through use of licenses, owners of intellectual assets grant other parties the right to access and make use of those assets. Licenses are thus the most common legal vehicle to facilitate the transfer of intellectual property.

The licensing process is thus the legal and commercial vehicle to manage use of intellectual assets. Licenses grant rights of use for intellectual assets. As a general matter, a range of factors should be incorporated into the effort to identify an appropriate royalty structure for an asset. The primary factors to be considered are the scope of the license and the payment structure.

Issues to be addressed with regard to the scope of a license for an intellectual asset include the issue of whether the license will be "exclusive" or "nonexclusive." An exclusive license makes the licensee the sole party to have the particular rights granted in the license. Thus, for example, if a license makes a licensee the exclusive user of an asset in a specific country, no other party will have the right to use that asset in the specified country. Exclusive licenses are of greater commercial value than nonexclusive licenses; thus, royalties associated with exclusive licenses are higher than those for nonexclusive licenses.

Other important aspects of the scope of a license are its duration and the range of rights granted. The duration of a license is its term, and generally the longer the term the more valuable the license and the higher the royalty. The range of rights includes the geographic scope of the license (e.g., rights in one country, a region, worldwide). It also includes the industries or fields of use in which the licensed asset can be used (e.g., a license might grant the right to use a technology only in a particular industry).

Another important factor pertaining to licensing is the ability of the licensee to redistribute the asset. Some licenses grant the licensee the right to authorize others to use the asset (e.g., the right to sublicense the asset to other parties). In general, the right to sublicense makes the license more valuable, thus requiring a greater royalty. Sublicensing rights also make it more difficult for the licensor to control the asset, thus introducing somewhat greater risk to the licensor.

Licensing decisions also require careful examination of the appropriate structure for the associated royalty payments. The first step in the development of a royalty payment structure is the issue of what the royalty amount will be. One common approach to establish an appropriate royalty level is to apply a royalty comparable to that applied to similar assets that are already under license. In addition to identifying the amount of the royalty, it is essential to define how the royalty will be calculated (e.g., a royalty paid per unit sold or one established as a percentage of total revenues from sales). The process for payment and collection of royalty must also be negotiated.

Online resources are now available to assist owners of intellectual property to develop licensing strategies. For example, services such as Royalty Source (www.royaltysource.com) can help owners of intellectual assets to develop appropriate royalties for the licensing of their assets. A service known as Patent ValuePredictor (www.patentvaluepredictor.com) helps owners of patents to identify the commercial value of those patents.

10.3 Auctions and exchanges

Buyers and sellers of intellectual assets increasingly meet in exchanges or auctions devoted to the transfer of such property. Auctions and exchanges have long been used to match buyers and sellers of traditional goods. Today, intellectual property is now commonly bought and sold in specialized marketplaces. Auction services such as Patent Auctions (www.patentauctions.com) provide a forum in which patent owners can auction their patents to others. The Patent & License Exchange (www.pl-x.com) provides an example of a trading exchange devoted to intellectual property. Systems such as IBM's intellectual property trading network (www.delphion.com) and the IP Network (www.ipnetwork.com) are examples of other intellectual asset exchanges. Even when transfer of intellectual assets is not fully consummated using an auction or exchange, use of those mechanisms facilitates intellectual asset transactions by helping buyers and sellers to find each other and by providing access to more comprehensive information, thus making the pricing process more efficient.

Expanded use of these trading centers makes the market for intellectual property more efficient. In that more efficient market, more knowledge assets can be commercially exploited, and the pricing associated with that use more accurately reflects supply and demand forces. In addition, these exchanges make it possible for assets that might have gone unused in the past to be more actively placed into commercial use. The trading system Patent Triage (www.patenttriage.com), for example, was established to find commercial applications for patents that were on the verge of being abandoned by their owners. Expect the trend toward greater use of auctions and trading exchanges for intellectual assets to continue in the future, as the number of developers and users of those assets increases.

10.4 Intellectual assets in corporate acquisitions and start-up enterprises

As the perceived economic value of intellectual assets increases, they play a growing role in decisions regarding corporate acquisitions and equity financing for start-up enterprises. In many instances, intellectual assets are the primary attraction motivating one company to acquire another or inspiring investors to make an equity investment in a new company. Investors commonly rely on their assessment of intellectual assets to influence their equity investment decisions. Businesses often choose merger and acquisition targets based, at least in part, on the intellectual asset portfolios of other companies.

The basic asset valuation models discussed previously in this chapter provide the fundamental tools used by investors and purchasers to support their investment and acquisition decisions. An important part of the "due diligence" review conducted by a potential buyer or investor is an examination of the strength of the intellectual asset portfolio of the target enterprise. As a general rule, that review should include an assessment of the enforceability of the ownership rights the target company holds for its intellectual assets. The more effectively enforceable those rights, the more valuable the target. Another important aspect of the review is an examination of the ability of the buyer or investor to make commercial use of the assets. For example, if a company is considering acquisition of a company with strong intellectual assets but learns that many of the employees who developed and work with those assets are likely to leave the company if there is an acquisition, the value of those assets is decreased. Mere existence of strong intellectual assets is not sufficient. Those assets must be presented in a manner that will facilitate optimal commercial use if they are to merit the maximum valuation.

In some situations, intellectual assets are the primary motivation for a business acquisition or an equity investment. A company may become an attractive acquisition target to an enterprise that would like to obtain access to intellectual assets controlled by the target company. Although licensing provides the most common means of access to intellectual assets of another organization, a merger or acquisition may make sense in instances when the assets desired are extensive or comprise a major portion of the target enterprise's total assets. This type of acquisition makes sense if the acquiring enterprise wants to make use of the target assets or if it believes that there will be competitive advantage from keeping the assets away from competitors.

Similarly, in some instances the appeal of specific intellectual assets controlled by a company may entice equity investors. Investors attracted by the commercial potential of those assets may invest in the entity that controls the assets, basing that investment decision on the anticipated potential economic return from future exploitation of the intellectual assets. Intellectual assets can thus be used as an important means of attracting equity funding. Prudent investors, however, are more attracted to those assets that have more readily quantifiable future value.

10.5 Securitization

An increasingly common practice applied to intellectual assets is use of those assets as collateral for loans. Just as land or other forms of traditional property are used as security to guarantee repayment of debts, so too can intellectual assets now be used as security. The use of intellectual property and other intellectual assets as security for loans is characterized as "securitization." Under this approach, intellectual assets are used as collateral to secure a loan.

Securitization is feasible only to the extent that lenders have confidence in their ability to value intellectual assets accurately and to liquidate those assets when necessary. Lenders are unwilling to accept intellectual assets as collateral if they are uncertain about the value of those assets or if they believe that it will be difficult for them to make use of the assets in the event of a default. Intellectual assets with highly speculative or inconsistent value are not popular as security.

In the event of default, intellectual assets that are pledged as collateral are claimed by the creditor. The process through which those assets are attached by the creditor is the same as that applied to conventional, tangible property. Creditors must follow their local laws to perfect the security interest they hold in the intellectual property, and after completing that process, the appropriate court will enforce their interest (In re Transportation Design & Technology [Bank. S.D. Cal. 1985]; *City Bank & Trust Co. v. Otto Fabric, Inc.* [D. Kan. 1988]; *Chesapeake Fiber Packaging Corp. v. Sebro Packaging Corp.* [D. Md. 1992]).

In bankruptcy, even those intellectual assets that were not pledged as collateral may be redistributed as part of the effort to liquidate assets. Just as traditional property is commonly sold to generate capital during bankruptcy proceedings, so too can intellectual assets be liquidated during bankruptcy reorganization for the benefit of creditors of the bankrupt company. Creditors seek to have their claims paid off during bankruptcy reorganization, and the trustees who manage that process now recognize that intellectual assets can be a profitable source of cash for use to repay creditors. Use of intellectual assets to pay off creditors during bankruptcy offers potential benefits for the creditors and for other parties who may have the opportunity to purchase those assets during bankruptcy sales. With the recent economic downturn, a growing number of companies now look at distressed businesses as a possible source of intellectual assets that can be acquired for substantially less than their market value.

10.6 Donating intellectual property

Another increasingly common intellectual asset management practice is the donation of the asset to a charitable organization. Just as traditional property can be donated to a charity, so too can patents and other forms of intellectual assets. Those donations of intellectual property can create income tax deductions, just as their traditional property counterparts do. Properly structured, donations of intellectual property create deductions that can be used to offset earned, taxable income.

To qualify for a tax deduction, donations of intellectual assets must meet several criteria that parallel those applied to traditional property. The asset must be donated to a charitable organization that is recognized by the tax regulations. The asset donated must have actual value, and the amount of that value must be established by a bona fide value assessor. The assessment of value must also be a current, timely estimate. In addition, the owner of the asset must relinquish all ownership rights to the nonprofit recipient. Total transfer of ownership and control of the asset is a critical element of charitable donations involving intellectual property.

Charitable donation of intellectual assets provides a potential means of extracting economic value from assets that do not fit the core business functions of their owners. Instead of leaving noncore intellectual assets unused, charitable donations permit the owners of those assets to derive a tax benefit (i.e., a deduction) based on the value of that donation. This process enables intellectual asset owners to recover some additional economic benefit from the assets even though they do not use those assets directly.

10.7 Abandonment

Under extreme conditions, some owners of intellectual assets simply abandon them. As is the case with traditional property, abandonment is simply the full relinquishment of all ownership rights for an asset. Abandonment is generally an action of last resort, since it creates no clear benefit for the asset owner other than the elimination of costs associated with maintenance of the ownership rights (e.g., cost of maintaining a patent). Generally, if there is any way to derive some positive value from an intellectual asset (e.g., through licensing or tax deductions associated with charitable donation of the asset), no matter how small that value may be, that alternative use of the asset is more advantageous than abandonment.

Note that abandonment is a more extreme action than any form of licensing. If, for example, an owner of a computer program makes that program widely available, at no charge (an Open Source distribution approach, for instance), that action does not constitute abandonment. Under Open Source and other open distribution models, the developer of the property generally continues to assert ownership over the property; however, the owner chooses to enter into liberal licensing arrangements. As long as an owner of an intellectual asset continues to assert ownership rights for that asset, the owner has not abandoned the asset.

10.8 Internal asset transfers

Intellectual property is now commonly transferred between affiliated enterprises (e.g., between subsidiaries of a common parent company) in order to minimize tax liability. Businesses that own intellectual property that is used in their operations now try to structure ownership and licensing to create an advantageous tax position. For instance, enterprises often transfer ownership of patents and other intellectual assets to a subsidiary located in a jurisdiction in which royalty revenues are either not taxed or are taxed at a favorable rate. That subsidiary can then license the property to other subsidiaries or the parent company. The income from those licenses is thus treated favorably, from an income tax perspective, and the affiliated entities that are paying the license fees can often treat those payments as business expenses and use those expenses to reduce taxable income. When tax treatment is favorable, this type of licensing structure involving affiliated business entities can result in a more advantageous tax position for the affiliated entities collectively than would be the case if the same entity owned and used the asset.

Transfer of ownership and rights of use of these assets to reduce tax liability is a common practice, since intellectual assets are easy to relocate. When implementing these internal transfers and licenses, it is important that the terms associated with those arrangements are reasonable and can be demonstrated to be the result of an "arm's length" transaction. If excessive use is made of special (i.e., particularly advantageous) arrangements between affiliated entities, government authorities can conclude that the arrangements are a sham and may refuse to grant the tax benefits for which the transaction would ordinarily qualify.

10.9 Auditing intellectual assets

A key first step to manage intellectual property and diverse intellectual assets is an audit of those assets. Intellectual asset audits have several stages. The first stage is the completion of an accurate and comprehensive inventory of the organization's intellectual assets. To create such an inventory, the organization should identify all traditional intellectual property (i.e., patents, copyrights, trademarks) used by the organization. This list includes both the intellectual property owned by the organization and the property owned by others and used by the organization. The inventory should then be expanded to include all identifiable intellectual assets (e.g., trade secrets, information, know-how, key human resources) that are not protected by traditional intellectual property law.

After completing the inventory of intellectual assets, the second stage of the audit involves identification of all of the ways in which the organization is currently using the various intellectual assets it possesses. In this phase of the audit, the organization strives to link its diverse intellectual assets with the products, services, and processes of the organization. This is the stage of the audit in which intellectual assets are identified with the specific goods, services, or business functions that they support.

Phase three involves analysis of which of those assets are of greatest importance to the organization. This step involves the identification of the core intellectual assets and the secondary intellectual assets. The categorization of each asset as core or secondary is based on the importance of the asset to a product, service, or function of the organization and on the importance of that product, or function, to the organization. For instance, an intellectual asset that provides essential support to a company's leading product or service, or to one of the company's key business activities, is a core asset. In contrast, an asset that provides minimal support to a minor product, service, or function is a noncore or secondary asset. This part of the audit identifies the intellectual assets that are of the greatest value for the organization and thus merit the most vigorous protection.

In phase four, the organization should review the strength of its control over the core intellectual assets. If the core asset is one that was developed by the organization, this stage of the audit should focus on determining whether the asset is adequately protected (e.g., have patent, copyright, or trademark rights been asserted, as appropriate). If the asset was not developed by the organization, a review should be conducted to ensure that the organization has obtained all the legal rights it requires in order to continue to use the asset (e.g., a valid licensing agreement). Based on this analysis, the organization should move to obtain, document, and enforce all legal rights necessary to ensure that the organization can continue to use those core intellectual assets.

In the final stage of the audit, all of the secondary intellectual assets should be examined. This review should focus on dividing the secondary assets into the following groups. One group includes the assets that may have value to another organization, and the second group consists of the intellectual assets that do not appear to have significant value to any identifiable organization. For the first group, there may be opportunities to license the asset to another organization and thus derive some future revenue. For the second group, revenue-generating licenses will be unlikely; however, there may still be ways to derive some benefit from the asset (e.g., a tax deduction associated with donation of the asset to a qualified charitable organization). The final stage of the audit provides the forum in which decisions addressing how to handle noncore intellectual assets should be made.

Audits of intellectual assets can be effectively performed by the organization that controls the assets or by independent third parties. The audit requires knowledge of the business and the technologies involved. It also requires expertise with regard to the legal aspects of ownership and use of intellectual assets, along with experience evaluating market forces associated with intellectual assets. Given these demands, audits of intellectual assets can be most effectively performed, for most companies, through judicious use of both internal and external resources. A comprehensive audit of intellectual assets sets the foundation for optimized management of all intellectual assets.

10.10 Antitrust and competition law constraints

Use of intellectual assets is constrained by antitrust and competition laws. That impact is expressed in two ways. Antitrust and competition laws restrict the ability of the owners of intellectual assets to use those assets to lessen commercial competition. Additionally, those laws assist the owners of intellectual assets in their efforts to manage access to, and use of, those assets by others.

Legal constraints are most rigorously enforced against owners of intellectual assets that are deemed to be "essential" for commercial use by other parties. Sometimes characterized as "essential facilities," access to these assets is necessary for the continued commercial viability of other users. For example, owners of intellectual assets that have large market share (e.g., Microsoft, Intel) often find that the law describes their assets as essential facilities (*Intergraph v. Intel*, Case No. 00-1368 [Fed. Cir 2001] and Case No. 00-1048 [11th Cir. 2001]; *U.S. v. Microsoft*, Case No. 00-5212 [D.C. Cir. 2001]). Those assets are defined in that manner, since other enterprises (e.g., computer equipment manufacturers) must have access to the assets or they will not be able to remain in business. Antitrust and competition laws are used to restrict the ability of owners of essential intellectual assets from exercising those ownership rights in ways that actually reduce competition or were intended to reduce competition (e.g., denying access to the assets or subjecting users of the assets to unreasonable licensing terms).

The requirements of antitrust law and competition law can affect the ability of intellectual asset owners to control access to their assets, even when the asset owner in question is not of the size of companies such as Microsoft or Intel. The key factor in this context is the extent to which the intellectual asset in question is essential for the continued commercial viability of another enterprise. If the asset is an essential facility, antitrust and competition laws are applied to ensure continued access to the facility, subject to reasonable terms. Accordingly, even relatively small companies that control an innovative intellectual asset that becomes widely applied may find that their asset has quickly become an essential facility, a condition that subjects them to more rigorous antitrust and competition laws oversight than they previously encountered.

Competition law also provides protection for owners of intellectual assets. It limits the ability of users of intellectual assets to the extent that those users try to access or apply the assets of others in unfair or unreasonable ways. An unfair competition claim can be raised by an owner of an intellectual asset against a user of that asset who makes use of the asset in ways that are inconsistent with the reasonable usage restrictions imposed by the asset owner. Examples of this application of competition law can be found in cases where owners of certain online content alleged that parties who operated content search systems were misusing the content in ways that resulted in unfair competition (*eBay v. Bidder's Edge*, 100 F. Supp. 2d 1058; *HOB Entertainment, Inc. v. Streambox, Inc.*, Case No. 00-10017 [C.D. Cal. filed Sept. 13, 2000]; *Register.com v. Verio*, 126 F. Supp. 2d 238).

Both the owners and the users of intellectual assets are subject to legal constraints intended to protect commercial competition. These constraints prevent an owner of essential intellectual assets from wielding rights of access to, and use of those assets in a manner that threatens competition. The constraints also prevent other parties from applying the assets in a manner that is inconsistent with the reasonable requirements imposed by the owner of the assets, to the potential detriment of the owner. As intellectual asset management strategies are developed and applied, owners and users of the assets should ensure that their activities do not have an adverse impact on the competitive environment in their markets. Of course, control of intellectual assets is intended to provide a competitive advantage, and efforts to develop and maintain such an advantage are entirely legal and appropriate. What is not legal or appropriate are efforts to use intellectual assets in ways that reduce competition.

10.11 Intellectual asset management strategies

Developers of intellectual property and other forms of intellectual assets should establish comprehensive management strategies for those assets. To develop effective management strategies for those assets, organizations that develop or use the assets must recognize the legal rights and responsibilities associated with intellectual assets. Rights of ownership and control associated with intellectual assets provide the mechanism to develop and implement asset management plans.

At the core of those management strategies should be a recognition that intellectual assets should be managed in substantially the same manner as traditional property. Intellectual assets are widespread and highly prized. In that setting, owners and users of those assets now apply the full range of ownership, access, and use models to those assets. Those models have long been applied to conventional property, and they are now widely applied to intellectual property and intellectual assets.

Expect the trend toward more diverse strategies for the management of intellectual assets to continue in the future. As those assets continue to become more widespread in the global economy, and their commercial value increases, there will be more incentive to apply the full range of property management strategies to those assets. Intellectual assets will continue to become increasingly mobile and liquid. Markets will become more adept at establishing value for those assets and more effective at transferring the assets as required to meet market demand.

Management decisions associated with intellectual assets should recognize both the offensive and the defensive aspects of those assets. Intellectual assets can be used offensively as a source of licensing revenues. They can also be used defensively to preserve competitive advantage by blocking or delaying competitors from entering into a market. Another defensive aspect of intellectual assets is their potential value as currency that can be used in "cross-licensing," where the asset owner licenses rights of use to the asset to another party who controls an intellectual asset that is important to the asset owner. When estimating the value of those assets and when developing strategies for use of the assets it is important to recognize that their value stems from both offensive and defensive applications.

Finally, it is important to recognize the full range of intellectual assets. There is often a tendency to focus on those assets that are eligible for traditional intellectual property law protection. This view grossly understates the value of intellectual assets. When identifying the assets available and when assessing their economic value, make sure that a broad perspective is applied. That perspective should be expansive enough to address conventional intellectual property and unconventional intellectual assets (including human resources, knowledge, and information). In some organizations, the unconventional intellectual assets have value that dwarfs the value of the traditional intellectual property. Failure to recognize the full scope of intellectual assets can lead to severe economic and competitive trauma for an organization.

10.12 Valuing intellectual assets in a networked setting

Assessment of the economic value of intellectual assets is a particularly challenging exercise when those assets are created, stored, and distributed on a computer network, such as the Internet. A network setting both enhances and diminishes the economic value of the intellectual assets contained on the network. Economic value is enhanced, since it is possible for users to find material that they sought, but could not obtain, in the past. The networks make it possible for users to access exceptional content that fulfills consumer demand. At the same time, however, computer networks can diminish the value of intellectual assets by diluting the supply of those assets. Networks make it possible for a wide range of suppliers to distribute their assets on the network. That dramatic increase in content supply can reduce the economic value of the more common content by providing multiple sources for that material, thus diluting the value of the material. By significantly increasing content supply, networks can also reduce content value by making it more difficult for network users to find the material they seek. If supply increases to a point that overwhelms network search and retrieval capabilities, then the value of the assets distributed through the network decreases over time, as consumers become less successful at finding the material they seek.

If the network expands while simultaneously preserving effective content management functions, it is likely that the existence of the network will, in general, increase the economic value of the assets made available through the network. The necessary content management functions include search, retrieval, and payment capabilities. To provide a setting in which consumers and suppliers of intellectual assets can derive the full potential economic benefit of networking, consumers must be able to find the material they seek efficiently and conveniently. Content search and retrieval capabilities are essential to meet that consumer requirement effectively. At the same time, content suppliers must be able to control content retrieval and purchasing functions, or the suppliers will have no incentive to make their assets available on the network.

At its current stage of development, the Internet is a laboratory hosting trials of different methods to provide the digital content management functions necessary to support an efficient online marketplace for intellectual assets. It remains unclear which economic models and commercial practices will provide the requisite level of both convenient and efficient consumer access and supplier security and confidence. One of the more promising of the current models is the "pay-per-use" approach, which customizes content to suit the specific demands of each consumer, while it also permits the supplier to assess and collect efficiently prices that are also customized to meet the actual conditions associated with each consumer's demand for the asset at the specific time of the transaction. The pay-per-use process enables each consumer to purchase the precise material that he or she wants and to purchase only the exact amount of the material that is desired (e.g., a single software application instead of a suite of applications or a single recorded song instead of an entire CD). The process enables suppliers to control access and payment securely, while assessing a price set exactly at the maximum level the specific consumer will accept under the market conditions that exist at the precise time of the sales transaction. Whether the pay-per-use model will, in fact, prevail in the future and what other commercial models will succeed as the networks mature remain open questions at this time. The answers to those questions will have a substantial impact on the intellectual asset valuation strategies of tomorrow.

Selected bibliography

Chesapeake Fiber Packaging Corp. v. Sebro Packaging Corp., 143 B.R. 360 (D. Md. 1992).

City Bank & Trust Co. v. Otto Fabric, Inc., 83 B.R. 780 (D. Kan. 1988).

eBay v. Bidder's Edge, 100 F. Supp. 2d 1058.

HOB Entertainment, Inc., et al. v. Streambox, Inc., Case No. 00-10017 (C.D. Cal. filed Sept. 13, 2000).

IBM IP Network, <www.delphion.com>.

Intergraph v. Intel, Case No. 00-1368 (D.C. Cir. 2001).

Intergraph v. Intel, Case No. 00-1048 (11th Cir. 2001).

In re Transportation Design & Technology, 48 B.R. 635 (Bank. S.D. Cal. 1985).

IP Network, <www.ipnetwork.com>.

Patent Auctions, <www.patentauction.com>.

Patent & License Exchange, <www.pl-x.com>.

Patent Man, <www.patentman.com>.

Patent Post, <www.patentpost.com>.

Patent Triage, <www.patenttriage.com>.

Patent Value Predictor, <www.patentvaluepredictor.com>.

Register.com v. Verio, 126 F. Supp. 2d 238.

Royalty Source, <www.royaltysource.com>.

U.S. v. Microsoft, Case No. 00-5212 (D.C. Cir. Jun. 28, 2001).

11

Principles for managing intellectual property and knowledge assets

We have discussed the different forms of digital property and knowledge assets that are widely recognized to have significant economic value. We have also reviewed some of the key opportunities and challenges currently associated with efforts to manage access to, and use of, those assets effectively in a networked environment. In this chapter, we will summarize some of the key strategic principles that should govern management of intellectual property and other knowledge assets in the future.

11.1 Understand the significance of computer networking for knowledge assets

The expansion of computer networking technology and applications has dramatically affected creation, distribution, and use of knowledge and intellectual assets. That expansion has made all network users into both creators and consumers of information, digital media content, and other forms of intellectual assets. In many ways, it is the movement toward ubiquitous computing and "always-on" networks that has led governments, businesses, and individuals to appreciate the economic value of a wide range of intellectual assets. The existence and expansion of those networks have also helped to enhance the economic value of the knowledge-based assets that are made accessible by the networks.

This direct link between computer networking and intellectual asset value is an important principle. It is a principle that will likely be even more critical in the future. Prudent developers, distributors, and users of intellectual assets will appreciate the connection between networking and intellectual asset value. They will understand that there is substantial economic value in increasing the number of users of, and uses for, the global computer network. Based on that recognition, an increasing number of content developers will strive to find ways to make their material more readily accessible online, provided that they are confident in the payment and security measures associated with online distribution. All providers and consumers of intellectual assets will eventually recognize that widely accessible computer networks profoundly enhance the value of those assets but also make management of the assets a greater challenge.

As developers of content move to place more of that material online, they will face pricing and timing challenges. In the online setting, supply of content is substantial; thus, there is significant competition in the marketplace, which tends to drive prices and profit margins down. As content providers move more of their inventory online, they often face the challenge of lower profits. To an extent, those reduced profit margins can be offset if market share increases. While the online supplier may earn less profit for each unit sold, if the total number of units sold increases enough, overall profitability can be preserved or even increased. A key strategic challenge facing suppliers of online content is the extent to which increased reliance on the online distribution system can create enough market share growth to offset reduced per unit profit margins and thus generate increased total profits.

Suppliers of knowledge assets in the networked setting should also apply an appropriate strategy for the timing and location of their market entry. Suppliers of intellectual assets must manage effectively the time and place of the release of their assets. For example, motion picture producers have developed a standard sequence for the commercial distribution of their works in different markets. Generally, theatrical release is followed by pay-per-view distribution, followed by cable television and video rental release, and ultimately broadcast television distribution. The producers apply this process of sequential release into different markets in an effort to maximize the total return from distribution of the asset in all markets. Similar strategies aimed at controlling the timing of the entry of intellectual assets into multiple markets should be developed for all forms of intellectual assets.

11.2 Recognize diverse forms of knowledge assets

In the past, discussion of intellectual property focused on the traditional forms of that property, as established under copyright, patent, and trademark law. Only those assets governed by basic intellectual property law were acknowledged to be true knowledge-based assets. Although we continue to direct substantial attention to traditional intellectual property, we now appreciate the fact that there are many different forms of knowledge, know-how, and information that may not qualify for basic intellectual property law protection, yet also have significant economic value. In the past, there was a tendency to value only those assets that could be protected under patent, copyright, or trademark laws. Today, it is widely recognized that there are many forms of intellectual assets that have significant economic value. In the future, we are likely to find that the value of knowledge assets other than traditional intellectual property is far greater than we ever believed it to be. The most successful creators and users of intellectual assets in the future will be those who appreciate the diversity of intellectual assets and those who effectively harness the commercial value of those assets.

The full range of legal principles and doctrines applies to management of intellectual assets created and distributed in the Internet environment. In the early days of the Internet, there was a tendency to focus on the application of traditional intellectual property laws to online content. Today, we recognize that effective management of intellectual assets in the networked environment requires a truly diverse range of laws and legal principles. Intellectual property law continues to play a major role in the development and use of intellectual assets, and it will continue to play a central role in the future. We now realize, however, that many additional laws and legal concepts, ranging from antitrust law to the law of torts, will also significantly affect management of knowledge and intellectual assets in the future, supplementing traditional intellectual property rights and complicating asset management efforts.

The different forms of knowledge and intellectual assets should be managed in different ways. Not all such assets can be managed through use of traditional intellectual property law principles. For example, a wide range of information and knowledge is treated as proprietary material under principles of trade secrets law or traditional property rights. In addition, some intellectual assets are best managed through more open access strategies, such as Open Source development processes or peer-to-peer file-sharing systems. Knowledge asset management strategy requires use of both proprietary and open access concepts in order to deal most effectively with the full range of intellectual assets.

Perhaps the most critical element of intellectual assets is people. People are the source of intellectual assets, and they should be included among those assets. Effective intellectual asset management thus requires effective management of human resources. In the future, this important connection between people and other critical intellectual assets will be more widely appreciated. The organizations that most fully appreciate this connection are likely to be the winners in the future competition to develop and make use of powerful intellectual assets.

11.3 Apply all available asset management tools

Effective management of intellectual property and knowledge assets requires application of the full range of management tools available for that property. Those management tools include application of legal rights, use of digital rights management technology, and use of economic incentives for both the developers and users of the property. Legal rights, technology, or economic forces alone will not provide maximum flexibility to support development and distribution of intellectual assets. Collectively, however, those three tools can enable owners of intellectual assets to exercise effective control over access and use of the assets.

The range of legal rights available to owners of intellectual assets for protection of their assets is increasing. In addition to traditional copyright, patent, and trademark rights, an increasing array of new legal concepts (e.g., database ownership rights, anticybersquatting rules) are now available to intellectual asset owners. A variety of property and commercial law principles (e.g., antitrust law, competition law, commercial transaction rules) are increasingly applied to manage access to intellectual assets. Effective intellectual asset management now requires an understanding of these different legal rights and a willingness to enforce all of them, as appropriate.

Developers of intellectual assets now commonly rely on digital rights management technology (e.g., encryption systems) to protect their content.

Technology provides a means of managing access and use that can be controlled entirely by the owner of the assets involved. Unlike legal strategies, which require assistance from legal institutions for enforcement, technology provides a "self-help" mechanism available for digital rights management. Prudent intellectual asset management strategies should include active use of digital rights management technology as a means of controlling assets. Recognize, however, that even the most effective technology, alone, will not provide long-term protection for the assets.

Economic incentives provide another important tool for control of intellectual asset use. Creative pricing and licensing terms provide an effective method through which owners of intellectual assets can manage more effectively access to, and use of, those assets by other parties. Use of both traditional and nontraditional economic strategies for these assets is an essential component of an effective overall management strategy.

11.4 Recognize transitory value of assets

Developers of intellectual assets should recognize that the commercial value of their assets is limited. In the past, there was a tendency to view intellectual property as more static asset, which could be defended and exploited for an extended period of time. Creators of content invested substantial resources in the development of the content, and then they invested significant resources to apply intellectual property law to protect their work. In today's networked environment, and for the future as well, the value of intellectual assets is far more transitory. It is unwise to assume that premium value for intellectual assets can be sustained for an extended period of time.

Rapid distribution and diffusion of knowledge and intellectual assets permit prompt duplication and modification of innovations. In that setting, valuable information and creative content move quickly through the marketplace. It is increasingly difficult to restrict that flow of ideas and knowledge. As a result, new information and assets are quickly embraced by other parties and promptly integrated into new applications. Competition for intellectual assets is fierce, and in that environment it is foolish to assume that any competitive advantage derived from new knowledge or creativity can be sustained for long. The create and defend approach to intellectual assets will become more and more difficult to apply in the future. As a result, the most successful developers of knowledge-based assets in the future will be those parties who build their commercial models upon the assumption that any competitive advantage they establish as a result of innovation will be a short-term advantage.

11.5 Make use of nontraditional commercial models

Intellectual property has traditionally been exploited through conventional licensing arrangements. Developers of intellectual assets commonly asserted their legal rights of ownership by requiring payment of royalty fees in exchange for rights of access and use for their creative property. Licensing continues to be an important commercial model for intellectual assets; however, in the future, alternative exploitation strategies will likely be increasingly important. Future creators of intellectual assets will be more open to economic models that provide for economic returns other than direct licensing revenues. They will use those alternative commercial strategies to supplement their traditional licensing revenue stream.

The Open Source software development process provides one example of a nontraditional commercial model for intellectual asset development. As we discussed previously, it is unclear which specific economic model or models will ultimately dominate the Open Source process; however, it is clear that the Open Source process will continue to be an important component of the overall software marketplace. The Open Source process is based on acceptance of economic benefits other than direct licensing revenues as compensation for the grant of access to the underlying code. In this way, Open Source presents an illustration of how nontraditional commercial models can develop into significant forces in the broader marketplace for intellectual assets. The Open Source model itself, along with other alternative economic strategies, will continue to play an important role in intellectual asset management in the future.

Asset owners will likely continue to experiment with many different commercial models and business strategies for their intellectual assets. Some content developers will choose to apply a "pay-per-play" economic model for access to their assets. With this approach, buyers pay only for the content they actually use. Other developers will apply an "all you can eat" model, where the user pays a set fee for unrestricted use of the material for a specific period of time. Both of these models are already in use for a variety of digital products, including computer software and recorded music. Most likely, we will see that developers of intellectual assets will make use of both of these models and probably many others as well, depending on the specific characteristics and goals associated with their various products in various markets. This diversity of economic models and commercial strategies will likely be a continuing attribute of the online marketplace for intellectual assets.

11.6 Recognize that different assets have different value

Not all intellectual assets have equal economic value. Effective management of those assets requires an appreciation of the fact that they have different values and should be used accordingly. The most valuable intellectual assets should be protected as rigorously as possible. The least valuable assets should be treated as commodities and their owners should try to use those assets in creative ways to maximize their total value. Examples of this approach include the provider of online content who makes that content widely available, at no charge, but who also provides more valuable content (e.g., premium versions of the free content) for a fee. A mistake commonly made by intellectual asset owners is to treat all of their assets as highly valuable assets and to apply legal and other management measures aggressively to restrict access to the assets. In the aggregate, this approach leads to a less than optimal asset management strategy, since it increases the asset management costs and foregoes potential value that could be derived from more creative use of the least valuable assets.

Another important aspect of effective intellectual asset valuation is the need to appreciate the different sources of asset value. Intellectual asset developers commonly recognize an asset's potential value as a source of licensing revenue and its value as a means of protecting a current competitive advantage (e.g., a patent for a technology that gives the owner's product a competitive advantage over products of its competitors). Asset developers, however, often fail to recognize the value of intellectual assets that make their operations more efficient and thus reduce operating costs. They also often fail to recognize the potential value of assets that could be exploited in industries other than their own through licensing. Asset owners also often fail to appreciate that some of their assets may have greater long-term value if they are widely distributed at minimal or no charge. The growth of the Open Source model illustrates how a more open distribution model can, under some circumstances, be economically effective.

11.7 Accommodate open access and broad collaboration

An important continuing trend for the foreseeable future will likely be increased reliance on collaboration for the development and use of intellectual assets. The Open Source software development process, peer-to-peer and other file-swapping systems, and distributed computing networks illustrate the trend toward more open access to knowledge assets, a trend likely to continue for the foreseeable future. In this environment, there will be greater sharing of access to intellectual assets. Traditionally, such access has been limited to rights of use but not rights of modification (e.g., the right to modify source code and create derivative works form the original code). In the future, more open environment rights of modification will be more widely granted, resulting in somewhat less control for the original developers of the intellectual assets.

Intellectual asset development will become a more interactive process over time. More of those assets will, in the future, be distributed as works in progress, subject to interaction between original developers and subsequent users. This process will not, of course, replace the traditional system of independent creation and controlled distribution. The collaborative development process will, however, become an important supplement to the traditional development process. The motivation for increased use of collaborative asset development will not be mandatory legal requirements but will instead be a conclusion reached by a growing number of content developers that the collaborative process provides a more effective means of creating and circulating innovative works. Market forces, not government actions, will most likely play the key role in encouraging the current trend toward collaborative development of intellectual assets.

Greater reliance on collaboration in the development of intellectual assets is also likely to surface through greater use of outsourcing and temporary development teams. Already, many organizations make active use of contractors and other independent developers to assist in the creation of intellectual assets. Some enterprises make use of online exchanges or marketplaces that match independent content creators with organizations that have content development needs, to work on specific development projects. In the future, it is likely that these and other development outsourcing processes will be highly popular, offering efficient and flexible methods for intellectual asset development.

11.8 Experiment with different distribution models

Expansion of digital networks and increased diversity of content accessible on these networks will dramatically alter the role of traditional intermediaries involved in the distribution of intellectual assets. The music and other traditional media industries are currently struggling to deal with the increased accessibility of their content through the Internet. No one has yet mastered the new digital media distribution process; however, it is already clear that the new process requires different actions and skills on the part of the various parties who participate in the process. Participants in the digital media distribution chain must be creative and flexible to ensure that they find ways to continue to add value to the distribution process.

In this evolving distribution system, intermediaries (whether they are music recording companies, print publishers, video producers, or any other media content provider) must experiment with different approaches to content distribution. Failure to take the risk of experimentation today can lead to disastrous long-term consequences if intermediaries opt not to adjust their roles in the distribution chain and, as a result, become outdated and irrelevant. Successful long-term strategies for digital asset management will require a willingness to experiment with unconventional distribution schemes. Those who are not prepared to bear the risk of experimentation run an even greater risk that they may ultimately find that they have been made irrelevant by a new distribution structure.

11.9 Understand the globalization of knowledge assets

We see today that intellectual property and knowledge assets are global in nature. Individuals and organizations around the world are now active creators and users of those assets. Increasingly, the world recognizes the economic value of those assets and the important role played by those assets in the process of economic development. To date, the United States has been the country that most clearly appreciated the economic potential that intellectual assets possess. In the future, many countries will appreciate that value and will move to encourage development and use of those assets in their countries. This expanded recognition of the importance of intellectual assets will significantly enhance the value of those assets, and it will dramatically increase the competition associated with creation and use of those assets.

In the future, the most successful creators and users of intellectual assets will recognize that the entire world is a potential marketplace for their assets. They will appreciate the value of making their assets available globally, and they will understand that the world also provides an important source of assets that they can use, including human resources. They will also recognize, however, that in the global marketplace for intellectual assets, competition will be fierce, and success will require the ability to remain innovative and to respond quickly to technological and market changes. The recognition of the globalization of intellectual assets will likely be one of the most important future trends in the field of technology, and those organizations that develop strategies to participate effectively in that trend will dramatically increase their prospects of commercial success.

11.10 Understand the trend toward empowerment of asset users

The networked environment empowers the users of the network by providing them with greater access to, and control over, digital content. Users of knowledge assets have greater choice as to the material they use and the ways in which they use those assets. As networks become more expansive and more powerful in the future, content users will have even greater flexibility and choice. Users of digital content will have more opportunities, in the future, to forge direct relationships with the developers of content, relying less on intermediaries for access to the material they seek. Creators and users of intellectual assets should both recognize this trend and understand that it will likely lead to greater power over intellectual assets in the hands of consumers of those assets than has existed in the past.

Along with this user empowerment, however, there will be increasing pressure on content users to understand and abide by the legitimate rights of the asset developers. Legal rights of ownership and commercial obligations to the asset developers will continue to exist, and as the users of those assets gain greater rights of use for the assets, they will be held responsible for compliance. A process of education is likely to be required to teach consumers both their rights and their duties with regard to use of intellectual assets. This education may prove to be less difficult than one might expect, as the line between content developers and content users becomes blurred with the rise in popularity of content distribution systems such as Open Source and peer to peer.

These more open systems promote collaborative content development, and in a collaborative setting content users become, at least in part, content developers. Widespread collaboration in digital asset development can, in effect, make more people into content creators, and in that environment it may be easier to teach all of the system participants the rights and obligations that come with content development and content access. Greater collaboration can give more people a stake in the development and enforcement of fair rights of ownership and use of intellectual assets. In that future environment, we may balance more effectively the rights of developers and consumers than we do today. Effective collaboration may well be the key to our future ability to derive the full potential economic benefit from the widespread access to diverse intellectual assets that global computer networks facilitate.

Appendix: Basic intellectual property forms

This appendix contains samples of several commonly used intellectual property forms provided by the U.S. government. These and other forms are available on-line at the U.S. Copyright Office Web site (www.copyright.gov/forms) and from the U.S. Patent and Trademark Office Web site (www.uspto.gov/web/forms). Both of those Web sites are very valuable resources for copyright, patent, and trademark protection information.

Copyright Office fees are subject to change.
For current fees, check the Copyright Office
website at *www.copyright.gov*, write the Copyright Office, or call (202) 707-3000.

FORM TX

For a Nondramatic Literary Work
UNITED STATES COPYRIGHT OFFICE

REGISTRATION NUMBER

TX	TXU

EFFECTIVE DATE OF REGISTRATION

Month	Day	Year

DO NOT WRITE ABOVE THIS LINE. IF YOU NEED MORE SPACE, USE A SEPARATE CONTINUATION SHEET.

1

TITLE OF THIS WORK ▼

PREVIOUS OR ALTERNATIVE TITLES ▼

PUBLICATION AS A CONTRIBUTION If this work was published as a contribution to a periodical, serial, or collection, give information about the collective work in which the contribution appeared.   **Title of Collective Work ▼**

If published in a periodical or serial give: Volume ▼   Number ▼   Issue Date ▼   On Pages ▼

2

a

NAME OF AUTHOR ▼

DATES OF BIRTH AND DEATH
Year Born ▼   Year Died ▼

Was this contribution to the work a "work made for hire"?
☐ Yes
☐ No

AUTHOR'S NATIONALITY OR DOMICILE
Name of Country
OR { Citizen of ▶
Domiciled in ▶

WAS THIS AUTHOR'S CONTRIBUTION TO THE WORK
Anonymous?   ☐ Yes ☐ No
Pseudonymous?   ☐ Yes ☐ No

If the answer to either of these questions is "Yes," see detailed instructions.

NATURE OF AUTHORSHIP Briefly describe nature of material created by this author in which copyright is claimed. ▼

NOTE

Under the law, the "author" of a "work made for hire" is generally the employer, not the employee (see instructions). For any part of this work that was "made for hire" check "Yes" in the space provided, give the employer (or other person for whom the work was prepared) as "Author" of that part, and leave the space for dates of birth and death blank.

b

NAME OF AUTHOR ▼

DATES OF BIRTH AND DEATH
Year Born ▼   Year Died ▼

Was this contribution to the work a "work made for hire"?
☐ Yes
☐ No

AUTHOR'S NATIONALITY OR DOMICILE
Name of Country
OR { Citizen of ▶
Domiciled in ▶

WAS THIS AUTHOR'S CONTRIBUTION TO THE WORK
Anonymous?   ☐ Yes ☐ No
Pseudonymous?   ☐ Yes ☐ No

If the answer to either of these questions is "Yes," see detailed instructions.

NATURE OF AUTHORSHIP Briefly describe nature of material created by this author in which copyright is claimed. ▼

c

NAME OF AUTHOR ▼

DATES OF BIRTH AND DEATH
Year Born ▼   Year Died ▼

Was this contribution to the work a "work made for hire"?
☐ Yes
☐ No

AUTHOR'S NATIONALITY OR DOMICILE
Name of Country
OR { Citizen of ▶
Domiciled in ▶

WAS THIS AUTHOR'S CONTRIBUTION TO THE WORK
Anonymous?   ☐ Yes ☐ No
Pseudonymous?   ☐ Yes ☐ No

If the answer to either of these questions is "Yes," see detailed instructions.

NATURE OF AUTHORSHIP Briefly describe nature of material created by this author in which copyright is claimed. ▼

3

a   **YEAR IN WHICH CREATION OF THIS WORK WAS COMPLETED**   This information must be given in all cases.
◀ Year

b   **DATE AND NATION OF FIRST PUBLICATION OF THIS PARTICULAR WORK**   Complete this information ONLY if this work has been published.   Month ▶   Day ▶   Year ▶   ◀ Nation

4

See instructions before completing this space.

COPYRIGHT CLAIMANT(S) Name and address must be given even if the claimant is the same as the author given in space 2. ▼

TRANSFER If the claimant(s) named here in space 4 is (are) different from the author(s) named in space 2, give a brief statement of how the claimant(s) obtained ownership of the copyright. ▼

APPLICATION RECEIVED

ONE DEPOSIT RECEIVED

TWO DEPOSITS RECEIVED

FUNDS RECEIVED

DO NOT WRITE HERE — OFFICE USE ONLY

MORE ON BACK ▶ • Complete all applicable spaces (numbers 5–9) on the reverse side of this page.
• See detailed instructions.   • Sign the form at line 8.

DO NOT WRITE HERE
Page 1 of _____ pages

EXAMINED BY _____	FORM TX
CHECKED BY _____	
☐ CORRESPONDENCE Yes	FOR COPYRIGHT OFFICE USE ONLY

DO NOT WRITE ABOVE THIS LINE. IF YOU NEED MORE SPACE, USE A SEPARATE CONTINUATION SHEET.

PREVIOUS REGISTRATION Has registration for this work, or for an earlier version of this work, already been made in the Copyright Office?
☐ Yes ☐ No If your answer is "Yes," why is another registration being sought? (Check appropriate box.) ▼
a. ☐ This is the first published edition of a work previously registered in unpublished form.
b. ☐ This is the first application submitted by this author as copyright claimant.
c. ☐ This is a changed version of the work, as shown by space 6 on this application.
If your answer is "Yes," give: **Previous Registration Number ▶**   **Year of Registration ▶**

5

DERIVATIVE WORK OR COMPILATION
Preexisting Material Identify any preexisting work or works that this work is based on or incorporates. ▼

a **6**

See instructions
before completing
this space.

Material Added to This Work Give a brief, general statement of the material that has been added to this work and in which copyright is claimed. ▼

b

DEPOSIT ACCOUNT If the registration fee is to be charged to a Deposit Account established in the Copyright Office, give name and number of Account.
Name ▼   **Account Number ▼**

a **7**

CORRESPONDENCE Give name and address to which correspondence about this application should be sent.   Name/Address/Apt/City/State/ZIP ▼

b

Area code and daytime telephone number ▶   Fax number ▶
Email ▶

CERTIFICATION* I, the undersigned, hereby certify that I am the
Check only one ▶
☐ author
☐ other copyright claimant
☐ owner of exclusive right(s)
☐ authorized agent of _____
of the work identified in this application and that the statements made   Name of author or other copyright claimant, or owner of exclusive right(s) ▲
by me in this application are correct to the best of my knowledge.

8

Typed or printed name and date ▼ If this application gives a date of publication in space 3, do not sign and submit it before that date.
_____ Date ▶ _____

Handwritten signature (X) ▼

X _____

Certificate will be mailed in window envelope to this address:	Name ▼	**YOU MUST:** • Complete all necessary spaces • Sign your application in space 8	**9**
	Number/Street/Apt ▼	**SEND ALL 3 ELEMENTS** **IN THE SAME PACKAGE:** 1. Application form 2. Nonrefundable filing fee in check or money order payable to *Register of Copyrights* 3. Deposit material	Fees are subject to change. For current fees, check the Copyright Office website at www.copyright.gov, write the Copyright Office, or call (202) 707-3000.
	City/State/ZIP ▼	**MAIL TO:** Library of Congress Copyright Office 101 Independence Avenue, S.E. Washington, D.C. 20559-6000	

*17 U.S.C. § 506(e): Any person who knowingly makes a false representation of a material fact in the application for copyright registration provided for by section 409, or in any written statement filed in connection with the application, shall be fined not more than $2,500.

Rev: June 2002—20,000   Web Rev: June 2002   ♻ Printed on recycled paper   U.S. Government Printing Office: 2000-461-113/20,021

Copyright Office fees are subject to change.
For current fees, check the Copyright Office
website at *www.copyright.gov*, write the Copyright Office, or call (202) 707-3000.

FORM PA

For a Work of the Performing Arts
UNITED STATES COPYRIGHT OFFICE

REGISTRATION NUMBER

PA	PAU

EFFECTIVE DATE OF REGISTRATION

Month	Day	Year

DO NOT WRITE ABOVE THIS LINE. IF YOU NEED MORE SPACE, USE A SEPARATE CONTINUATION SHEET.

1

TITLE OF THIS WORK ▼

PREVIOUS OR ALTERNATIVE TITLES ▼

NATURE OF THIS WORK ▼ See instructions

2

a

NAME OF AUTHOR ▼

DATES OF BIRTH AND DEATH
Year Born ▼   Year Died ▼

Was this contribution to the work a "work made for hire"?
☐ Yes
☐ No

AUTHOR'S NATIONALITY OR DOMICILE
Name of Country
OR { Citizen of _____
Domiciled in _____

WAS THIS AUTHOR'S CONTRIBUTION TO THE WORK
Anonymous?   ☐ Yes ☐ No
Pseudonymous? ☐ Yes ☐ No
If the answer to either of these questions is "Yes," see detailed instructions.

NATURE OF AUTHORSHIP Briefly describe nature of material created by this author in which copyright is claimed. ▼

NOTE

Under the law, the "author" of a "work made for hire" is generally the employer, not the employee (see instructions). For any part of this work that was "made for hire" check "Yes" in the space provided, give the employer (or other person for whom the work was prepared) as "Author" of that part, and leave the space for dates of birth and death blank.

b

NAME OF AUTHOR ▼

DATES OF BIRTH AND DEATH
Year Born ▼   Year Died ▼

Was this contribution to the work a "work made for hire"?
☐ Yes
☐ No

AUTHOR'S NATIONALITY OR DOMICILE
Name of Country
OR { Citizen of _____
Domiciled in _____

WAS THIS AUTHOR'S CONTRIBUTION TO THE WORK
Anonymous?   ☐ Yes ☐ No
Pseudonymous? ☐ Yes ☐ No
If the answer to either of these questions is "Yes," see detailed instructions.

NATURE OF AUTHORSHIP Briefly describe nature of material created by this author in which copyright is claimed. ▼

c

NAME OF AUTHOR ▼

DATES OF BIRTH AND DEATH
Year Born ▼   Year Died ▼

Was this contribution to the work a "work made for hire"?
☐ Yes
☐ No

AUTHOR'S NATIONALITY OR DOMICILE
Name of Country
OR { Citizen of _____
Domiciled in _____

WAS THIS AUTHOR'S CONTRIBUTION TO THE WORK
Anonymous?   ☐ Yes ☐ No
Pseudonymous? ☐ Yes ☐ No
If the answer to either of these questions is "Yes," see detailed instructions.

NATURE OF AUTHORSHIP Briefly describe nature of material created by this author in which copyright is claimed. ▼

3

a

YEAR IN WHICH CREATION OF THIS WORK WAS COMPLETED This information must be given Year in all cases.

b

DATE AND NATION OF FIRST PUBLICATION OF THIS PARTICULAR WORK
Complete this information ONLY if this work has been published.
Month _____ Day _____ Year _____
Nation _____

4

See instructions before completing this space.

COPYRIGHT CLAIMANT(S) Name and address must be given even if the claimant is the same as the author given in space 2. ▼

TRANSFER If the claimant(s) named here in space 4 is (are) different from the author(s) named in space 2, give a brief statement of how the claimant(s) obtained ownership of the copyright. ▼

APPLICATION RECEIVED

ONE DEPOSIT RECEIVED

TWO DEPOSITS RECEIVED

FUNDS RECEIVED

DO NOT WRITE HERE OFFICE USE ONLY

MORE ON BACK ▶
• Complete all applicable spaces (numbers 5-9) on the reverse side of this page.
• See detailed instructions.
• Sign the form at line 8.

DO NOT WRITE HERE
Page 1 of _____ pages

EXAMINED BY	FORM PA
CHECKED BY	
☐ CORRESPONDENCE Yes	FOR COPYRIGHT OFFICE USE ONLY

DO NOT WRITE ABOVE THIS LINE. IF YOU NEED MORE SPACE, USE A SEPARATE CONTINUATION SHEET.

PREVIOUS REGISTRATION Has registration for this work, or for an earlier version of this work, already been made in the Copyright Office?

☐ Yes ☐ No If your answer is "Yes," why is another registration being sought? (Check appropriate box.) ▼ If your answer is No, do not check box A, B, or C.

a. ☐ This is the first published edition of a work previously registered in unpublished form.

b. ☐ This is the first application submitted by this author as copyright claimant.

c. ☐ This is a changed version of the work, as shown by space 6 on this application.

If your answer is "Yes," give: **Previous Registration Number** ▼   **Year of Registration** ▼

5

DERIVATIVE WORK OR COMPILATION Complete both space 6a and 6b for a derivative work; complete only 6b for a compilation.

Preexisting Material Identify any preexisting work or works that this work is based on or incorporates. ▼

a

6

Material Added to This Work Give a brief, general statement of the material that has been added to this work and in which copyright is claimed. ▼

b

See instructions before completing this space.

DEPOSIT ACCOUNT If the registration fee is to be charged to a Deposit Account established in the Copyright Office, give name and number of Account.

Name ▼   **Account Number** ▼

a

7

CORRESPONDENCE Give name and address to which correspondence about this application should be sent. Name/Address/Apt/City/State/ZIP ▼

b

Area code and daytime telephone number   (   )   Fax number   (   )

Email

CERTIFICATION* I, the undersigned, hereby certify that I am the

Check only one ▶
☐ author
☐ other copyright claimant
☐ owner of exclusive right(s)
☐ authorized agent of

Name of author or other copyright claimant, or owner of exclusive right(s) ▲

of the work identified in this application and that the statements made by me in this application are correct to the best of my knowledge.

8

Typed or printed name and date ▼ If this application gives a date of publication in space 3, do not sign and submit it before that date.

Date

Handwritten signature (X) ▼

☞   x _____

Certificate will be mailed in window envelope to this address:	Name ▼	**YOU MUST:** • Complete all necessary spaces • Sign your application in space 8 **SEND ALL 3 ELEMENTS IN THE SAME PACKAGE:** 1. Application form 2. Nonrefundable filing fee in check or money order payable to *Register of Copyrights* 3. Deposit material **MAIL TO:** Library of Congress Copyright Office 101 Independence Avenue, S.E. Washington, D.C. 20559-6000	**9**
	Number/Street/Apt ▼		
	City/State/ZIP ▼		

*17 U.S.C. § 506(e): Any person who knowingly makes a false representation of a material fact in the application for copyright registration provided for by section 409, or in any written statement filed in connection with the application, shall be fined not more than $2,500.

Rev: June 2002—20,000   Web Rev: June 2002   ⊕ Printed on recycled paper   U.S. Government Printing Office: 2000-461-113/20,021

Copyright Office fees are subject to change. For current fees, check the Copyright Office website at *www.copyright.gov*, write the Copyright Office, or call (202) 707-3000.

FORM VA
For a Work of the Visual Arts
UNITED STATES COPYRIGHT OFFICE

REGISTRATION NUMBER

VA   VAU

EFFECTIVE DATE OF REGISTRATION

Month   Day   Year

DO NOT WRITE ABOVE THIS LINE. IF YOU NEED MORE SPACE, USE A SEPARATE CONTINUATION SHEET.

1 Title of This Work ▼   NATURE OF THIS WORK ▼ See Instructions

Previous or Alternative Titles ▼

Publication as a Contribution If this work was published as a contribution to a periodical, serial, or collection, give information about the collective work in which the contribution appeared. Title of Collective Work ▼

If published in a periodical or serial give: Volume ▼   Number ▼   Issue Date ▼   On Pages ▼

2 a NAME OF AUTHOR ▼   DATES OF BIRTH AND DEATH
Year Born ▼   Year Died ▼

NOTE
Under the law, the "author" of a "work made for hire" is generally the employer, not the employee (see instructions). For any part of this work that was "made for hire" check "Yes" in the space provided, give the employer (or other person for whom the work was prepared) as "Author" of that part, and leave the space for dates of birth and death blank.

Was this contribution to the work a "work made for hire"? ☐ Yes ☐ No

Author's Nationality or Domicile Name of Country OR { Citizen of _____ Domiciled in _____

Was This Author's Contribution to the Work
Anonymous? ☐ Yes ☐ No
Pseudonymous? ☐ Yes ☐ No
If the answer to either of these questions is "Yes," see detailed instructions.

Nature of Authorship Check appropriate box(es). See Instructions
☐ 3-Dimensional sculpture ☐ Map ☐ Technical drawing
☐ 2-Dimensional artwork ☐ Photograph ☐ Text
☐ Reproduction of work of art ☐ Jewelry design ☐ Architectural work

b Name of Author ▼   Dates of Birth and Death
Year Born ▼   Year Died ▼

Was this contribution to the work a "work made for hire"? ☐ Yes ☐ No

Author's Nationality or Domicile Name of Country OR { Citizen of _____ Domiciled in _____

Was This Author's Contribution to the Work
Anonymous? ☐ Yes ☐ No
Pseudonymous? ☐ Yes ☐ No
If the answer to either of these questions is "Yes," see detailed instructions.

Nature of Authorship Check appropriate box(es). See instructions
☐ 3-Dimensional sculpture ☐ Map ☐ Technical drawing
☐ 2-Dimensional artwork ☐ Photograph ☐ Text
☐ Reproduction of work of art ☐ Jewelry design ☐ Architectural work

3 a Year in Which Creation of This Work Was Completed   This information must be given   Year in all cases.
b Date and Nation of First Publication of This Particular Work Complete this information ONLY if this work has been published. Month _____ Day _____ Year _____ Nation

4 COPYRIGHT CLAIMANT(S) Name and address must be given even if the claimant is the same as the author given in space 2. ▼

Transfer If the claimant(s) named here in space 4 is (are) different from the author(s) named in space 2, give a brief statement of how the claimant(s) obtained ownership of the copyright. ▼

See instructions before completing this space.

APPLICATION RECEIVED
ONE DEPOSIT RECEIVED
TWO DEPOSITS RECEIVED
FUNDS RECEIVED
DO NOT WRITE HERE OFFICE USE ONLY

MORE ON BACK ▶ • Complete all applicable spaces (numbers 5-9) on the reverse side of this page.
• See detailed instructions. • Sign the form at line 8.

DO NOT WRITE HERE
Page 1 of _____ pages

EXAMINED BY _____   **FORM VA**

CHECKED BY _____

☐ CORRESPONDENCE
Yes

FOR COPYRIGHT OFFICE USE ONLY

DO NOT WRITE ABOVE THIS LINE. IF YOU NEED MORE SPACE, USE A SEPARATE CONTINUATION SHEET.

5 **PREVIOUS REGISTRATION** Has registration for this work, or for an earlier version of this work, already been made in the Copyright Office?

☐ Yes ☐ No If your answer is "Yes," why is another registration being sought? (Check appropriate box.) ▼

a. ☐ This is the first published edition of a work previously registered in unpublished form.

b. ☐ This is the first application submitted by this author as copyright claimant.

c. ☐ This is a changed version of the work, as shown by space 6 on this application.

If your answer is "Yes," give: **Previous Registration Number** ▼   **Year of Registration** ▼

6 **DERIVATIVE WORK OR COMPILATION** Complete both space 6a and 6b for a derivative work; complete only 6b for a compilation.

a. **Preexisting Material** Identify any preexisting work or works that this work is based on or incorporates. ▼

a   See instructions before completing this space.

b. **Material Added to This Work** Give a brief, general statement of the material that has been added to this work and in which copyright is claimed. ▼

b

7 **DEPOSIT ACCOUNT** If the registration fee is to be charged to a Deposit Account established in the Copyright Office, give name and number of Account.

Name ▼   **Account Number** ▼

a

CORRESPONDENCE Give name and address to which correspondence about this application should be sent. Name/Address/Apt/City/State/ZIP ▼

b

Area code and daytime telephone number ( )   Fax number ( )

Email

8 **CERTIFICATION*** I, the undersigned, hereby certify that I am the

check only one ▶ ☐ author
☐ other copyright claimant
☐ owner of exclusive right(s)
☐ authorized agent of _____
Name of author or other copyright claimant, or owner of exclusive right(s) ▲

of the work identified in this application and that the statements made by me in this application are correct to the best of my knowledge.

Typed or printed name and date ▼ If this application gives a date of publication in space 3, do not sign and submit it before that date.

Date _____

Handwritten signature (X) ▼

X _____

9 Certificate will be mailed in window envelope to this address:

Name ▼

Number/Street/Apt ▼

City/State/ZIP ▼

YOU MUST:
• Complete all necessary spaces
• Sign your application in space 8
SEND ALL 3 ELEMENTS IN THE SAME PACKAGE:
1. Application form
2. Nonrefundable filing fee in check or money order payable to Register of Copyrights
3. Deposit material
MAIL TO:
Library of Congress
Copyright Office
101 Independence Avenue, S.E.
Washington, D.C. 20559-6000

*17 U.S.C. § 506(e): Any person who knowingly makes a false representation of a material fact in the application for copyright registration provided for by section 409, or in any written statement filed in connection with the application, shall be fined not more than $2,500.

Rev: June 2002—20,000 Web Rev: June 2002 ♻ Printed on recycled paper   U.S. Government Printing Office: 2000-461-113/20,021

Copyright Office fees are subject to change.
For current fees, check the Copyright Office
website at *www.copyright.gov*, write the Copyright Office, or call (202) 707-3000.

FORM SE

For a Serial
UNITED STATES COPYRIGHT OFFICE

REGISTRATION NUMBER

_____

U

EFFECTIVE DATE OF REGISTRATION

_____

Month   Day   Year

DO NOT WRITE ABOVE THIS LINE. IF YOU NEED MORE SPACE, USE A SEPARATE CONTINUATION SHEET.

1

TITLE OF THIS SERIAL ▼

Volume ▼   Number ▼   Date on Copies ▼   Frequency of Publication ▼

PREVIOUS OR ALTERNATIVE TITLES ▼

2

a

NAME OF AUTHOR ▼

DATES OF BIRTH AND DEATH
Year Born ▼   Year Died ▼

Was this contribution to the work a "work made for hire"?
☐ Yes
☐ No

AUTHOR'S NATIONALITY OR DOMICILE
Name of Country
OR { Citizen of ▶_____
Domiciled in▶_____

WAS THIS AUTHOR'S CONTRIBUTION TO THE WORK
Anonymous?   ☐ Yes ☐ No
Pseudonymous?   ☐ Yes ☐ No
If the answer to either of these questions is "Yes," see detailed instructions.

NATURE OF AUTHORSHIP Briefly describe nature of material created by this author in which copyright is claimed. ▼
☐ Collective Work   Other:

NOTE

Under the law, the "author" of a "work made for hire" is generally the employer, not the employee (see instructions). For any part of this work that was "made for hire" check "Yes" in the space provided, give the employer (or other person for whom the work was prepared) as "Author" of that part, and leave the space for dates of birth and death blank.

b

NAME OF AUTHOR ▼

DATES OF BIRTH AND DEATH
Year Born ▼   Year Died ▼

Was this contribution to the work a "work made for hire"?
☐ Yes
☐ No

AUTHOR'S NATIONALITY OR DOMICILE
Name of Country
OR { Citizen of ▶_____
Domiciled in▶_____

WAS THIS AUTHOR'S CONTRIBUTION TO THE WORK
Anonymous?   ☐ Yes ☐ No
Pseudonymous?   ☐ Yes ☐ No
If the answer to either of these questions is "Yes," see detailed instructions.

NATURE OF AUTHORSHIP Briefly describe nature of material created by this author in which copyright is claimed. ▼
☐ Collective Work   Other:

c

NAME OF AUTHOR ▼

DATES OF BIRTH AND DEATH
Year Born ▼   Year Died ▼

Was this contribution to the work a "work made for hire"?
☐ Yes
☐ No

AUTHOR'S NATIONALITY OR DOMICILE
Name of Country
OR { Citizen of ▶_____
Domiciled in▶_____

WAS THIS AUTHOR'S CONTRIBUTION TO THE WORK
Anonymous?   ☐ Yes ☐ No
Pseudonymous?   ☐ Yes ☐ No
If the answer to either of these questions is "Yes," see detailed instructions.

NATURE OF AUTHORSHIP Briefly describe nature of material created by this author in which copyright is claimed. ▼
☐ Collective Work   Other:

3

a

YEAR IN WHICH CREATION OF THIS ISSUE WAS COMPLETED This information must be given in all cases.
◀Year

b

DATE AND NATION OF FIRST PUBLICATION OF THIS PARTICULAR ISSUE
Complete this information ONLY if this work has been published.
Month ▶_____ Day▶_____ Year▶_____
◀ Nation

4

See instructions before completing this space.

COPYRIGHT CLAIMANT(S) Name and address must be given even if the claimant is the same as the author given in space 2. ▼

TRANSFER If the claimant(s) named here in space 4 is (are) different from the author(s) named in space 2, give a brief statement of how the claimant(s) obtained ownership of the copyright. ▼

APPLICATION RECEIVED

ONE DEPOSIT RECEIVED

TWO DEPOSITS RECEIVED

REMITTANCE NUMBER AND DATE

DO NOT WRITE HERE OFFICE USE ONLY

MORE ON BACK ▶ • Complete all applicable spaces (numbers 5-11) on the reverse side of this page.
• See detailed instructions.   • Sign the form at line 10.

DO NOT WRITE HERE
Page 1 of_____ pages

EXAMINED BY		FORM SE
CHECKED BY		
☐	CORRESPONDENCE Yes	FOR COPYRIGHT OFFICE USE ONLY

DO NOT WRITE ABOVE THIS LINE. IF YOU NEED MORE SPACE, USE A SEPARATE CONTINUATION SHEET.

PREVIOUS REGISTRATION Has registration for this work, or for an earlier version of this work, already been made in the Copyright Office?

☐ Yes   ☐ No   If your answer is "Yes," why is another registration being sought? (Check appropriate box.) ▼

a. ☐ This is the first published edition of a work previously registered in unpublished form.

b. ☐ This is the first application submitted by this author as copyright claimant.

c. ☐ This is a changed version of the work, as shown by space 6 on this application.

If your answer is "Yes," give: **Previous Registration Number** ▼   **Year of Registration** ▼

5

DERIVATIVE WORK OR COMPILATION Complete both space 6a and 6b for a derivative work; complete only 6b for a compilation.

Preexisting Material Identify any preexisting work or works that this work is based on or incorporates. ▼   **a**

6

See instructions before completing this space.

Material Added to This Work Give a brief, general statement of the material that has been added to this work and in which copyright is claimed. ▼   **b**

DEPOSIT ACCOUNT If the registration fee is to be charged to a Deposit Account established in the Copyright Office, give name and number of Account.

Name ▼   **Account Number** ▼   **a**

7

CORRESPONDENCE Give name and address to which correspondence about this application should be sent.   Name / Address / Apt / City / State / ZIP ▼   **b**

Area code and daytime telephone number ▶   Fax number ▶

Email ▶

CERTIFICATION* I, the undersigned, hereby certify that I am the

Check only one ▶
☐ author
☐ other copyright claimant
☐ owner of exclusive right(s)
☐ authorized agent of _____

of the work identified in this application and that the statements made by me in this application are correct to the best of my knowledge.

Name of author or other copyright claimant, or owner of exclusive right(s) ▲

8

Typed or printed name and date ▼ If this application gives a date of publication in space 3, do not sign and submit it before that date.

_____   Date ▶

Handwritten signature (X) ▼

X _

Certificate will be mailed in window envelope to this address:	Name ▼	YOU MUST: • Complete all necessary spaces • Sign your application in space 8	**9**
	Number/Street/Apt ▼	SEND ALL 3 ELEMENTS IN THE SAME PACKAGE: 1. Application form 2. Nonrefundable filing fee in check or money order payable to *Register of Copyrights* 3. Deposit material	
	City/State/ZIP ▼	MAIL TO: Library of Congress Copyright Office 101 Independence Avenue, S.E. Washington, D.C. 20559-6000	

*17 U.S.C. § 506(e): Any person who knowingly makes a false representation of a material fact in the application for copyright registration provided for by section 409, or in any written statement filed in connection with the application, shall be fined not more than $2,500.

Rev. June 2002—20,000   Web Rev. June 2002   ♻ Printed on recycled paper   U.S. Government Printing Office: 2000-461-113/20,021

Copyright Office fees are subject to change. For current fees, check the Copyright Office website at *www.copyright.gov*, write the Copyright Office, or call (202) 707-3000.

FORM RE
For Renewal of a Work
UNITED STATES COPYRIGHT OFFICE

REGISTRATION NUMBER

EFFECTIVE DATE OF RENEWAL REGISTRATION

Month	Day	Year

DO NOT WRITE ABOVE THIS LINE. IF YOU NEED MORE SPACE, USE A SEPARATE CONTINUATION SHEET (FORM RE/CON).

1 RENEWAL CLAIMANT(S), ADDRESS(ES), AND STATEMENT OF CLAIM ▼ (See Instructions)

a
Name ....
Address ....
Claiming as ....
(Use appropriate statement from instructions)

b
Name ....
Address ....
Claiming as ....

c
Name ....
Address ....
Claiming as ....

2 TITLE OF WORK IN WHICH RENEWAL IS CLAIMED ▼

RENEWABLE MATTER ▼

PUBLICATION AS A CONTRIBUTION If this work was published as a contribution to a periodical, serial, or other composite work, give information about the collective work in which the contribution appeared.   **Title of Collective Work ▼**

If published in a periodical or serial give: **Volume ▼**   **Number ▼**   **Issue Date ▼**

3 AUTHOR(S) OF RENEWABLE MATTER ▼ (If any author is deceased, give month, day, and year of death.)

4 ORIGINAL REGISTRATION NUMBER ▼   ORIGINAL COPYRIGHT CLAIMANT ▼

ORIGINAL DATE OF COPYRIGHT
If the original registration for this work was made in published form, give: If the original registration for this work was made in unpublished form, give:
DATE OF PUBLICATION: _____   **OR**   DATE OF REGISTRATION: _____
(Month)   (Day)   (Year)   (Month)   (Day)   (Year)

MORE ON BACK ➤   • Complete all applicable spaces (numbers 5–8) on the reverse side of this page.   DO NOT WRITE HERE
• See detailed instructions.   • Sign the form at space 7.   Page 1 of _____ pages

RENEWAL APPLICATION RECEIVED	FORM RE
CORRESPONDENCE ❏ YES	
EXAMINED BY	FOR
CHECKED BY	COPYRIGHT OFFICE
FUNDS RECEIVED	USE ONLY

DO NOT WRITE ABOVE THIS LINE. IF YOU NEED MORE SPACE, USE A SEPARATE CONTINUATION SHEET (FORM RE/CON).

RENEWAL FOR GROUP OF WORKS BY SAME AUTHOR: To make a single registration for a group of works by the same individual author published as contributions to periodicals (see instructions), give full information about each contribution. If more space is needed, request continuation sheet (Form RE/CON).

5

a
Title of Contribution: ....
Title of Periodical: .... Vol: .... No: .... Issue Date: ....
Date of Publication: .... Registration Number: ....
(Month)   (Day)   (Year)

b
Title of Contribution: ....
Title of Periodical: .... Vol: .... No: .... Issue Date: ....
Date of Publication: .... Registration Number: ....
(Month)   (Day)   (Year)

c
Title of Contribution: ....
Title of Periodical: .... Vol: .... No: .... Issue Date: ....
Date of Publication: .... Registration Number: ....
(Month)   (Day)   (Year)

d
Title of Contribution: ....
Title of Periodical: .... Vol: .... No: .... Issue Date: ....
Date of Publication: .... Registration Number: ....
(Month)   (Day)   (Year)

DEPOSIT ACCOUNT: If the registration fee is to be charged to a Deposit Account established in the Copyright Office, give name and number of Account.

Name _____

Account Number _____

Area code and daytime telephone number ▶ _____

CORRESPONDENCE: Give name and address to which correspondence about this application should be sent.

Name _____

Address _____ (Apt)

(City)   (State)   (ZIP)

Fax number ▶ _____ Email Address ▶ _____

6

CERTIFICATION* I, the undersigned, hereby certify that I am the: (Check one)
❏ renewal claimant   ❏ duly authorized agent of _____
(Name of renewal claimant ▲)
of the work identified in this application and that the statements made by me in this application are correct to the best of my knowledge.

Typed or printed name ▼ _____ Date ▼ _____

Handwritten signature (X) ▼ _____

7

Certificate will be mailed in window envelope to this address:

Name ▼ _____

Number/Street/Apt ▼ _____

City/State/ZIP ▼ _____

YOU MUST:
• Complete all necessary spaces
• Sign your application in space 7

SEND ALL ELEMENTS IN THE SAME PACKAGE:
1. Application form
2. Nonrefundable filing fee in check or money order payable to *Register of Copyrights*

MAIL TO:
Library of Congress
Copyright Office
101 Independence Avenue, S.E.
Washington, D.C. 20559-6000

Fees are subject to change. For current fees, check the Copyright Office website at www.copyright.gov, write the Copyright Office, or call (202) 707-3000.

8

*17 U.S.C. § 506(e): Any person who knowingly makes a false representation of a material fact in the application for copyright registration provided for by section 409, or in any written statement filed in connection with the application, shall be fined not more than $2,500.

Rev: December 2002—20,000 Web Rev: December 2002 ♻ Printed on recycled paper

U.S. Government Printing Office: 2000-461-113/20,021

⊘ Form MW ⊘

Form MW should be used for registration of a claim for protection in a mask work which is fixed in a semiconductor chip product, by or under the authority of the owner of the mask work.

A "mask work" is a series of related images, however fixed or encoded, having or representing the predetermined, three-dimensional pattern of metallic, insulating, or semiconductor material present or removed from the layers of a semiconductor chip product, and in which the relation of the images to one another is such that each image has the pattern of the surface of one form of the semiconductor chip product. To be protected, a mask work must be original. If the mask work consists of designs that are staple, commonplace, or familiar in the semiconductor industry, any variation or combination of such designs must, considered as a whole, be original to be protected. In no case does protection for a mask work extend to any idea, procedure, process, system, method of operation, concept, discovery, or the like embodied or illustrated in a mask work. Nor is a design protectible if it is dictated by a particular electronic function or is one of only a few available design choices that will accomplish that function.

Protection for a mask work begins on the date on which the mask work is registered, or the date of first commercial exploitation anywhere in the world, whichever occurs first, and lasts for a term of 10 years. During such time, the owner has the exclusive rights to: 1) reproduce the mask work by any means; 2) import or distribute semiconductor chip products embodying the mask work; and 3) authorize or induce others to reproduce, to import, or to distribute.

Statutory protection for a mask work that has been commercially exploited anywhere in the world is terminated if application for registration of a claim for protection is not made within 2 years after the date of first commercial exploitation.

DEPOSIT TO ACCOMPANY APPLICATION: The Act requires the deposit of identifying material as specified in the Copyright Office Regulations. (37 C.F.R. §211). For further information, request Circular 96 Part 211.

MASK WORK PROTECTION DIFFERENT FROM COPYRIGHT: Both the Copyright Act and the Semiconductor Chip Protection Act are admin-istered by the Copyright Office and involve protection for intellectual property. However, they differ from each other in most respects, including term, ownership, eligibility, scope of exclusive rights, limitations on exclusive rights, remedies, and registration procedures.

FOR FORMS OR INFORMATION: For information, write or call the Copyright Office, Library of Congress, Washington, D.C. 20559-6000, (202) 707-3000. For forms, call (202) 707-9100 or download them from *www.copyright.gov*. You may photocopy blank application forms; **however,** photocopies of Form MW submitted to the Copyright Office must be clear, legible, on a good grade of 8 1/2-inch by 11-inch (preferably blue) paper, suitable for automatic feeding through a photocopier. **Forms not meeting these requirements will be returned.**

PRIVACY ACT ADVISORY STATEMENT. Required by the Privacy Act of 1974 (P.L. 93-579):
The authority for requesting this information is title 17 U.S.C., sec. 908, which provides for mandatory registration of claims to mask work protection. Furnishing of the information is voluntary, but if the information is not furnished, it is probable that registration will be refused. Unless a judicial appeal should result in an order compelling registration, any inchoate rights in the mask work would be forfeited at the expiration of 2 years from first commercial exploitation.
The principal uses of the requested information are the examination of the application for registration to determine compliance with legal requirements and the establishment and maintenance of a public record of claims of protection.
Other routine uses include public inspection and copying, preparation of public indexes, preparation of public catalogs of mask work registrations, and preparation of search reports upon request.
NOTE: No other advisory statement will be given in connection with this application. Please keep this statement and refer to it if we communicate with you regarding this application.

MASK WORK NOTICE: The owner of a protected mask work may affix a notice to the mask work or to the semiconductor chip product embodying the mask work in such a way as to give reasonable notice of such protection. The notice consists of two elements: 1) the words "mask work," the symbol *M*, or the letter M in a circle μ ; and 2) the name of the owner or owners of the mask work or an abbreviation by which the name is recognized or is generally known. The affixation of a notice is not a condition of protection under the law, but provides certain benefits.

▬ SPACE BY SPACE INSTRUCTIONS ▬

1 Space 1: TITLE.
Every work submitted for registration must be given a title for purposes of cataloging and identification. This title may include the name of the semiconductor chip product in which the mask work is embodied, e.g., "ASTRA 2014," "Memory Cell 5522," or "Register X22."

2 Space 2: NATURE OF DEPOSIT.
Give a short description of the object deposited as identifying material, e.g., "chips plus seven of nine acetate layers," "acetate color overlay sheets," or "composite plot."

3, 4, 5 Spaces 3, 4, and 5: INFORMATION ABOUT CURRENT OWNER(S).
The owner of a mask work is: 1) the person who created the mask work; 2) the legal representative of that person if that person is deceased or under a legal incapacity; 3) the employer for whom a person created the mask work within the scope of his or her employment; or 4) the party to whom all the rights of such a person, employer or representative are transferred. Give the name(s) and address(es) of the current owner(s) of the mask work which is the subject matter of this application. Use a continuation sheet if additional space is needed.
Give the citizenship or domicile of the current owner in space 4.
If the current owner is not the person who created the mask work that is the subject matter of this application, check the appropriate box in space 5 to explain how the owner acquired the right to claim protection in this mask work. NOTE: If the current owner is a company or organization, one of the boxes must be checked.

6 Space 6: DATE AND NATION OF FIRST COMMERCIAL EXPLOITATION.
To "commercially exploit" a mask work is to distribute to the public for commercial purposes a semiconductor chip product embodying the mask work. The offering to sell or transfer a semiconductor chip product is a commercial exploitation only when the offer is in writing and occurs after the fixation of a mask work in a semiconductor chip product.
If this mask work has been commercially exploited anywhere in the world, give the exact date (month, day and year) and the nation of first commercial exploitation. If the work has not yet been commercially exploited, leave this space blank.

7 Space 7: CITIZENSHIP OR DOMICILE OF OWNER AT TIME OF FIRST COMMERCIAL EXPLOITATION.
Eligibility for protection may depend on the nationality or domicile of the owner of a commercially exploited mask work at the time of first commercial exploitation. Complete this space if the mask work which is the subject of this application was commercially exploited, and if the nationality or domicile of the owner at the time of first commercial exploitation is different from that given in space 4.

8 Space 8: NATURE OF CONTRIBUTION.
Mask works generally contain preexisting material that is common in the semiconductor industry. Such material is not protectible. However, if staple designs are combined in a way that is original, the new authorship may be protected. Further, portions of a work that may have been previously commercially exploited or previously registered for protection may not be included in the claim.
Give a brief, general statement that describes the new protectible contribution that is the basis of this claim. This statement may, if appropriate, refer to any previous mask work upon which the mask work being registered is based, as an aid in distinguishing the new contribution from the preexisting material. NOTE: Protection does not extend to the functions of the semiconductor chip product.

9, 10, 11, 12 Spaces 9, 10, 11, and 12: CORRESPONDENCE. In space 9, give the name, address, email, and daytime telephone number of the contact person if further information about this claim is needed.
DEPOSIT ACCOUNT. Complete space 10 if an existing deposit account is to be charged for the filing fee.
CERTIFICATION. Give the handwritten signature of a person authorized to certify the facts asserted in this application. **The application must be signed. ADDRESS FOR RETURN OF CERTIFICATE.** The name and address must be completed legibly. The certificate will be mailed in a window envelope.

Copyright Office fees are subject to change. For current fees, check the Copyright Office website at *www.copyright.gov*, write the Copyright Office, or call (202) 707-3000.

FORM MW
For Mask Works
UNITED STATES COPYRIGHT OFFICE
REGISTRATION NUMBER

MW
EFFECTIVE DATE OF REGISTRATION

Month	Day	Year

APPLICATION RECEIVED

DEPOSIT RECEIVED

SEE SPACE-BY-SPACE INSTRUCTIONS ON OTHER SIDE. DO NOT WRITE ABOVE THIS LINE. IF YOU NEED MORE SPACE, USE FORM MW/CON.

1. TITLE OF THIS WORK

2. NATURE OF DEPOSIT

3. NAME AND ADDRESS OF CURRENT OWNER(S)

4. CITIZENSHIP OR DOMICILE OF CURRENT OWNER(S)
Citizen of: _____
or
Domiciled in: _____

5. DERIVATION OF OWNERSHIP: If the person who created the mask work which is subject matter of this application is NOT named as the owner, check one: (Note: If a company or organization is named as the current owner, one of the following boxes *must* be checked.)
☐ a. The owner is the employer of a person who created such mask work within the scope of his/her employment.
☐ b. The owner has acquired the rights by transfer from the creator, employer or representative.
☐ c. The owner is the legal representative of the deceased or legally incapacitated creator.

6. DATE AND NATION OF FIRST COMMERCIAL EXPLOITATION
Month _____ Day _____ Year _____
Nation _____

7. CITIZENSHIP OR DOMICILE OF OWNER AT THE TIME OF FIRST COMMERCIAL EXPLOITATION (See instructions)
Citizen of: _____
Domiciled in: _____

8. NATURE OF CONTRIBUTION: Mask works generally contain designs that are staple, commonplace, or familiar in the semiconductor industry, or are variations of such designs, or are variations of designs that have been previously commercially exploited or previously registered for protection. Describe the new, original contribution in this mask work for which statutory protection is sought: _____

9. CONTACT PERSON FOR CORRESPONDENCE ABOUT THIS CLAIM
Name: _____
Daytime telephone number: ( ) _____
Address (if other than given at space 12): _____

FAX: ( ) _____ Email: _____

10. DEPOSIT ACCOUNT
Name of Account: _____

Account Number: _____

11. CERTIFICATION: I, the undersigned, hereby certify that I have the authority to submit this application and that the statements made herein are correct to the best of my knowledge.*

HANDWRITTEN SIGNATURE (X) _____
(This application MUST be signed.)
TYPED SIGNATURE _____

12.
Certificate will be mailed in window envelope to this address:

Name

Number/Street/Apt Number

City/State/ZIP

PLEASE BE SURE THAT YOU HAVE:
• Signed the application at space 11.
• Enclosed a check or money order for the nonrefundable filing fee of $75, payable to Register of Copyrights.
• Enclosed deposit, application, and fee.

MAIL TO:
Library of Congress
Copyright Office, Department MW
101 Independence Avenue, S.E.
Washington, D.C. 20540

*Any person who knowingly makes a false representation of a material fact in the application for registration as provided in 18 USC 1001 shall be fined not more than $10,000.

Rev: June 2002—20,000 Web Rev: June 2002 ℗ Printed on recycled paper

U.S. Government Printing Office: 2000-461-113/20,021

PTO Form 1478 (Rev 9/98)
OMB No. 0651-0009 (Exp. 08/31/2004)

Trademark/Service Mark Application, Principal Register, with Declaration

eTEAS - Version 1.24a: 01/01/2002

Each field name links to the relevant section of the "HELP" instructions that will appear at the bottom of the screen. Fields containing the symbol "*" **must** be completed; all other relevant fields should be completed if the information is known. If there are multiple owners or if the goods and/or services are classified in more than one class, click on the Form Wizard.
Note: ☐ check here if you do not want the scrolling help to be automatically shown at the bottom of the screen.

i

Important: ONCE AN APPLICATION IS SUBMITTED ELECTRONICALLY, THE OFFICE WILL IMMEDIATELY PROVIDE THE SENDER WITH AN ELECTRONIC ACKNOWLEDGMENT OF RECEIPT OF THE APPLICATION. Please contact the Office within 24 hours of transmission (or by the next business day) if you do not receive this acknowledgment. Contact Points:

For general trademark information, please telephone the Trademark Assistance Center, at 703-308-9000. For automated status information on an application that has an assigned serial number, please telephone 703-305-8747, or use http://tarr.uspto.gov.

If you need help in resolving technical glitches, you can e-mail us at PrinTEAS@uspto.gov. Please include your telephone number in your Email, so we can talk to you directly, if necessary.

Instructions

To file the application electronically, please complete the following steps:

1. Fill out all mandatory fields, which are noted by an * symbol.
2. Validate the form, using the "button" at the end of the form. If there are errors, go back to step 1.
3. If the desired signatory is not available to sign the application, to forward the form to the signatory, use either the Text Form option or the Download Portable Form option. Both options are available from the Validation Page. However, to use the Text Form option, you must first answer YES to Form Wizard Question #9. When you receive the application back, return to step 2.
4. Use the Pay/Submit button at the bottom of the Validation Screen. This will allow you to choose from 3 different payment methods: credit card, automated deposit account, or electronic funds transfer. After accessing the proper screen for payment, and making the appropriate entries, you will receive a confirmation screen if your transmission is successful. This screen will say SUCCESS! and will provide your assigned serial number.
5. You will receive an e-mail acknowledgement of your submission, which will repeat the assigned serial number and provide a summary of your submission.

Applicant Information

Please use the Wizard if there are multiple applicants.

* Name	[If an individual, use following format: Last Name, First Name, Middle Initial/Name]	

Entity Type: Click on the **one** appropriate circle to indicate the applicant's entity type and enter the corresponding information.

○ Individual	Country of Citizenship	
○ Corporation	State or Country of Incorporation	
○ Partnership	State or Country Where Organized	
	Name and Citizenship of all General Partners	
○ Other	Specify Entity Type	
	State or Country Where Organized	
* Address	* Street Address	
	* City	
	State	Select State
		If not listed above, please select 'OTHER' and specify here:
	* Country	Select Country
		If not listed above, please select 'OTHER' and specify here:
	Zip/Postal Code	
Phone Number		
Fax Number		
Internet E-Mail Address		□ Check here to authorize the USPTO to communicate with the applicant or its representative via e-mail. NOTE: While the application may list an e-mail address for the applicant, applicant's attorney, and/or applicant's domestic representative, **only one** e-mail address may be used for correspondence, in accordance with Office policy. The applicant must keep this address current in the Office's records.

Mark Information

Before the USPTO can register your mark, we must know exactly what it is. You can display a mark in one of two formats:
(1) typed; or (2) stylized or design. When you click on one of the two circles below, and follow the relevant instructions, the program will create a separate page that displays your mark.

WARNING: AFTER SEARCHING THE USPTO DATABASE, EVEN IF YOU THINK THE RESULTS ARE "O.K.," DO NOT ASSUME THAT YOUR MARK CAN BE REGISTERED AT THE USPTO. AFTER YOU FILE AN APPLICATION, THE USPTO MUST DO ITS OWN SEARCH AND OTHER REVIEW, AND MIGHT REFUSE TO REGISTER YOUR MARK.

* Mark	○ **Typed Format**	Click on this circle if you wish to register a word(s), letter(s), and/or number(s) in a format that can be reproduced using a typewriter. Also, only the following common punctuation marks and symbols are acceptable in a typed drawing (any other symbol requires a stylized format): . ? " - ; ( ) % $ @ + , ! ' : / & # * = [ ] Enter the mark here: NOTE: The mark **must** be entered in ALL upper case letters, regardless of how you actually use the mark. E.g., MONEYWISE, **not** MoneyWise.
	○ **Stylized or Design Format**	Click on this circle if you wish to register a stylized word(s), letter(s), number(s), and/or a design. The design may also include words. Click on the 'Browse' button to select GIF or JPG image file from your local drive that shows the complete, overall mark (i.e., the stylized representation of the words, e.g., or if a design that also includes words, the image of the "composite" mark, NOT just the design element). Do NOT submit a color image. [Browse...] For a stylized word(s) or letter(s), or a design that also includes a word (s), enter the LITERAL element only of the mark here:
Additional Statement	This section is for the entry of various statements that may pertain to the mark. In no case must you enter any of these statements for the application to be accepted for filing (although you may be required to add a statement(s) to the record during the actual prosecution of the application). To select a statement, check the box and enter the specific information relevant to your mark. The following are the texts of the most commonly asserted statements:	

Additional Statement section:

☐ DISCLAIMER: "No claim is made to the exclusive right to use [____] apart from the mark as shown."

☐ STIPPLING AS A FEATURE OF THE MARK: "The stippling is a feature of the mark."

☐ STIPPLING FOR SHADING: "The stippling is for shading purposes only."

☐ PRIOR REGISTRATION(S): "Applicant claims ownership of U.S. Registration Number(s) [____]."

☐ DESCRIPTION OF THE MARK: "The mark consists of [____]"

☐ TRANSLATION: "The foreign wording in the mark translates into English as [____]."

☐ TRANSLITERATION: "The non-Latin character(s) in the mark transliterate into [____], and this means [____] in English."

☐ §2(f), based on Use: "The mark has become distinctive of the goods/services through the applicant's substantially exclusive and continuous use in commerce for at least the five years immediately before the date of this statement."

☐ §2(f), based on Prior Registration(s): "The mark has become distinctive of the goods/services as evidenced by the ownership on the Principal Register for the same mark for related goods or services of U.S. Registration No(s). [____]."

☐ §2(f), IN PART, based on Use: "[____] has become distinctive of the goods/services through the applicant's substantially exclusive and continuous use in commerce for at least the five years immediately before the date of this statement."

☐ §2(f), IN PART, based on Prior Registration(s): "[____] has become distinctive of the goods/services as evidenced by the ownership on the Principal Register for the same mark for related goods or services of U.S. Registration No(s).

☐ **NAME(S), PORTRAIT(S), SIGNATURE(S) OF INDIVIDUAL(S):**

   ○ "The name(s), portrait(s), and/or signature(s) shown in the mark identifies [   ], whose consent(s) to register will be submitted.

   ○ "The name(s), portrait(s), and/or signature(s) shown in the mark does not identify a particular living individual.

☐ **USE OF THE MARK IN ANOTHER FORM:** "The mark was first used anywhere in a different form other than that sought to be registered on [   ], and in commerce on [   ]."

☐ **CONCURRENT USE:** Enter the appropriate concurrent use information, e.g., specify the goods and the geographic area for which registration is sought.

BASIS FOR FILING AND GOODS AND/OR SERVICES INFORMATION

Applicant requests registration of the trademark/service mark identified above with the Patent and Trademark Office on the Principal Register established by the Act of July 5, 1946 (15 U.S.C. §1051 et seq.) for the following Class(es) and Goods and/or Services, and checks the basis that covers those specific Goods or Services. More than one basis may be selected, but do **NOT** claim both §§1(a) and 1(b) for the identical goods or services in one application.

☐ **Section 1(a), Use in Commerce: Applicant is using or is using through a related company the mark in commerce on or in connection with the below identified goods and/or services. 15 U.S.C. § 1051 (a), as amended. Applicant attaches one specimen for** *each class* **showing the mark as used in commerce on or in connection with any item in the class of listed goods and/or services. If filing electronically, applicant must attach a JPG or GIF specimen image file for each international class, regardless of whether the mark itself is in a typed drawing format or is in a stylized format or a design. Unlike the mark image file, a specimen image file may be in color (i.e., if color is being claimed as a feature of the mark, then the specimen image should show use of the actual color(s) claimed).**

Specimen Image File

   Click on the 'Browse' button to select GIF or JPG image file that contains the specimen from applicant's local drive.

   [   ] [Browse...]

Describe what the specimen submitted consists of:

International Class	[ ] If known, enter class number 001 - 045, A, B, or 200	
* Listing of Goods and/or Services *USPTO Goods/Services Manual*		
Date of First Use of Mark Anywhere	at least as early as: [ ] MM/DD/YYYY	

Date of First Use of the Mark in Commerce	
	MM/DD/YYYY

☐ **Section 1(b), Intent to Use:** Applicant has a bona fide intention to use or use through a related company the mark in commerce on or in connection with the goods and/or services identified below (15 U.S.C. §1051(b)).

International Class	If known, enter class number 001 - 045, A, B, or 200
* Listing of Goods and/or Services *USPTO Goods/Services Manual*	

☐ **Section 44(d), Priority based on foreign filing:** Applicant has a bona fide intention to use the mark in commerce on or in connection with the goods and/or services identified below, and asserts a claim of priority based upon a foreign application in accordance with 15 U.S.C. §1126(d).

International Class	If known, enter class number 001 - 045, A, B, or 200
* Listing of Goods and/or Services *USPTO Goods/Services Manual*	
Country of Foreign Filing	Select Country — If not listed above, please select 'OTHER' and specify here:
Foreign Application Number	NOTE:If possible, enter no more than 12 characters. Eliminate all spaces and non-alphanumeric characters.
Date of Foreign Filing	MM/DD/YYYY

☐ **Section 44(e), Based on Foreign Registration:** Applicant has a bona fide intention to use the mark in commerce on or in connection with the above identified goods and/or services, and will submit a certification or certified copy of the foreign registration before the application may proceed to registration, in accordance with 15 U. S.C. 1126(e), as amended.

International Class	If known, enter class number 001 - 045, A, B, or 200
* Listing of Goods and/or Services *USPTO Goods/Services Manual*	
Country of Foreign Registration	Select Country — If not listed above, please select 'OTHER' and specify here:
Foreign Registration Number	

		NOTE:If possible, enter no more than 12 characters. Eliminate all spaces and non-alphanumeric characters.
Foreign Registration Date		MM/DD/YYYY
Renewal Date for Foreign Registration		MM/DD/YYYY
Expiration Date of Foreign Registration		MM/DD/YYYY

☐ Check here if an attorney is filing this application on behalf of applicant(s). Otherwise, click on Domestic Representative to continue.

Attorney Information		
Correspondent Attorney Name		
Individual Attorney Docket/Reference Number		
Other Appointed Attorney(s)		
Attorney Address	Street Address	
	City	
	State	Select State ▼ If not listed above, please select 'OTHER' and specify here:
	Country	Select Country ▼ If not listed above, please select 'OTHER' and specify here:
	Zip/Postal Code	
Firm Name		
Phone Number		
FAX Number		
Internet E-Mail Address		☐ Check here to authorize the USPTO to communicate with the applicant or its representative via e-mail. NOTE: While the application may list an e-mail address for the applicant, applicant's attorney, and/or applicant's domestic representative, **only one** e-mail address may be used for correspondence, in accordance with Office policy. The applicant must keep this address current in the Office's records.

☐ Check here if the applicant has appointed a Domestic Representative. **A Domestic Representative is REQUIRED if the applicant's address is outside the United States.** Otherwise, click on Fee Information to continue.

Domestic Representative

The applicant **must** appoint a Domestic Representative if the applicant's address is outside the United States. The following is hereby appointed applicant's representative upon whom notice or process in the proceedings affecting the mark may be served.

Representative's Name		
Address	Street Address	
	City	
	State	Select State ▾
		If not listed above, please select 'OTHER' and specify here:
	Zip Code	
Firm Name		
Phone Number		
FAX Number		
Internet E-Mail Address		☐ Check here to authorize the USPTO to communicate with the applicant or its representative via e-mail. NOTE: While the application may list an e-mail address for the applicant, applicant's attorney, and/or applicant's domestic representative, **only one** e-mail address may be used for correspondence, in accordance with Office policy. The applicant must keep this address current in the Office's records.

Fee Information

Number of Classes Paid   `1` ▾

Note: The total fee is computed based on the Number of Classes in which the goods and/or services associated with the mark are classified.

$ `325`   = **Number of Classes Paid x $325 (per class)**

* Amount $ `____`

NOTE: TEAS has changed its payment options and procedures. Three options (credit card, automated deposit account (New!), and Electronic Funds Transfer (New!)) will now appear after clicking on the PAY/SUBMIT button, which is available on the bottom of the Validation Page after completing and validating the application form.

Declaration

The undersigned, being hereby warned that willful false statements and the like so made are punishable by fine or imprisonment, or both, under 18 U.S.C. §1001, and that such willful false statements may jeopardize the validity of the application or any resulting registration, declares that he/she is properly authorized to execute this application on behalf of the applicant; he/she believes the applicant to be the owner of the trademark/service mark sought to be registered, or, if the application is being filed under 15 U.S.C. §1051(b), he/she believes applicant to be entitled to use such mark in commerce; to the best of his/her knowledge and belief no other person, firm, corporation, or association has the right to use the mark in commerce, either in the identical form thereof or in such near resemblance thereto as to be likely, when used on or in connection with the goods/services of such other person, to cause confusion, or to cause mistake, or to deceive; and that all statements made of his/her own knowledge are true; and that all statements made on information and belief are believed to be true.

Electronic Signature

The application will not be "signed" in the sense of a traditional paper document. To verify the contents of the application, the signatory must enter any combination of alpha/numeric characters that has been specifically adopted to serve the function of the signature, preceded and followed by the forward slash (/) symbol. Acceptable "signatures" could include: /john doe/; /jd/; and /123-4567/. The application may still be verified to check for missing information or errors even if the signature and date signed fields are left blank.

Signature [   ]   Date Signed [   ]
                                                MM/DD/YYYY

Signatory's Name [   ]

Signatory's Position [   ]

Click on the desired action:

The "Validate Form" function allows you to run an automated check to ensure that all mandatory fields have been completed. You will receive an "error" message if you have not filled in one of the five (5) fields that are considered "minimum filing requirements" under the Trademark Law Treaty Implementation Act of 1998. For other fields that the USPTO believes are important, but not mandatory, you will receive a "warning" message if the field is left blank. This warning is a courtesy, if non-completion was merely an oversight. If you so choose, you may by-pass that "warning" message and validate the form (however, you cannot by-pass an "error" message).

[Validate Form]   [Reset Form]

Note: To either print the completed application, in whole or in part, download and save the validated application, or electronically submit the application to the USPTO, click on the Validate Form button.

Privacy Policy Statement

PTO/SB/01 (10-01)
Approved for use through 10/31/2002. OMB 0651-0032
U.S. Patent and Trademark Office; U.S. DEPARTMENT OF COMMERCE
Under the Paperwork Reduction Act of 1995, no persons are required to respond to a collection of information unless it contains a valid OMB control number.

DECLARATION FOR UTILITY OR DESIGN PATENT APPLICATION (37 CFR 1.63)	**Attorney Docket Number**	
	First Named Inventor	
	COMPLETE IF KNOWN	
	Application Number	
☐ Declaration Submitted with Initial Filing **OR** ☐ Declaration Submitted after Initial Filing (surcharge (37 CFR 1.16 (e)) required)	Filing Date	
	Art Unit	
	Examiner Name	

As the below named inventor, I hereby declare that:

My residence, mailing address, and citizenship are as stated below next to my name.

I believe I am the original and first inventor of the subject matter which is claimed and for which a patent is sought on the invention entitled:

(Title of the Invention)

the specification of which

☐ is attached hereto

OR

☐ was filed on (MM/DD/YYYY) _____ as United States Application Number or PCT International

Application Number _____ and was amended on (MM/DD/YYYY) _____ (if applicable).

I hereby state that I have reviewed and understand the contents of the above identified specification, including the claims, as amended by any amendment specifically referred to above.

I acknowledge the duty to disclose information which is material to patentability as defined in 37 CFR 1.56, including for continuation-in-part applications, material information which became available between the filing date of the prior application and the national or PCT international filing date of the continuation-in-part application.

I hereby claim foreign priority benefits under 35 U.S.C. 119(a)-(d) or (f), or 365(b) of any foreign application(s) for patent, inventor's or plant breeder's rights certificate(s), or 365(a) of any PCT international application which designated at least one country other than the United States of America, listed below and have also identified below, by checking the box, any foreign application for patent, inventor's or plant breeder's rights certificate(s), or any PCT international application having a filing date before that of the application on which priority is claimed.

Prior Foreign Application Number(s)	Country	Foreign Filing Date (MM/DD/YYYY)	Priority Not Claimed	Certified Copy Attached?	
				YES	NO
			☐	☐	☐
			☐	☐	☐
			☐	☐	☐

☐ Additional foreign application numbers are listed on a supplemental priority data sheet PTO/SB/02B attached hereto:

[Page 1 of 2]

Burden Hour Statement: This form is estimated to take 21 minutes to complete. Time will vary depending upon the needs of the individual case. Any comments on the amount of time you are required to complete this form should be sent to the Chief Information Officer, U.S. Patent and Trademark Office, Washington, DC 20231. DO NOT SEND FEES OR COMPLETED FORMS TO THIS ADDRESS. SEND TO: Assistant Commissioner for Patents, Washington, DC 20231.

PTO/SB/01 (10-01)
Approved for use through 10/31/2002. OMB 0651-0032
U.S. Patent and Trademark Office; U.S. DEPARTMENT OF COMMERCE
Under the Paperwork Reduction Act of 1995, no persons are required to respond to a collection of information unless it contains a valid OMB control number.

DECLARATION — Utility or Design Patent Application

Direct all correspondence to: ☐	Customer Number or Bar Code Label		OR ☐ Correspondence address below

Name

Address

City	State	ZIP

Country	Telephone	Fax

I hereby declare that all statements made herein of my own knowledge are true and that all statements made on information and belief are believed to be true; and further that these statements were made with the knowledge that willful false statements and the like so made are punishable by fine or imprisonment, or both, under 18 U.S.C. 1001 and that such willful false statements may jeopardize the validity of the application or any patent issued thereon.

NAME OF SOLE OR FIRST INVENTOR : ☐ A petition has been filed for this unsigned inventor

Given Name (first and middle [if any])	Family Name or Surname

Inventor's Signature	Date

Residence: City	State	Country	Citizenship

Mailing Address

City	State	ZIP	Country

NAME OF SECOND INVENTOR: ☐ A petition has been filed for this unsigned inventor

Given Name (first and middle [if any])	Family Name or Surname

Inventor's Signature	Date

Residence: City	State	Country	Citizenship

Mailing Address

City	State	ZIP	Country

☐ Additional inventors are being named on the ____supplemental Additional Inventor(s) sheet(s) PTO/SB/02A attached hereto.

[Page 2 of 2]

Please type a plus sign (+) inside this box ➡ ☐

PTO/SB/05 (03-01)
Approved for use through 10/31/2002. OMB 0651-0032
U.S. Patent and Trademark Office; U.S. DEPARTMENT OF COMMERCE
Under the Paperwork Reduction Act of 1995, no persons are required to respond to a collection of information unless it displays a valid OMB control number.

UTILITY PATENT APPLICATION TRANSMITTAL

(Only for new nonprovisional applications under 37 CFR 1.53(b))

Attorney Docket No.	
First Inventor	
Title	
Express Mail Label No.	

APPLICATION ELEMENTS
See MPEP chapter 600 concerning utility patent application contents.

1. ☐ Fee Transmittal Form (e.g., PTO/SB/17)
(Submit an original and a duplicate for fee processing)

2. ☐ Applicant claims small entity status. See 37 CFR 1.27.

3. ☐ Specification [Total Pages ____]
(preferred arrangement set forth below)
- Descriptive title of the invention
- Cross Reference to Related Applications
- Statement Regarding Fed sponsored R & D
- Reference to sequence listing, a table, or a computer program listing appendix
- Background of the Invention
- Brief Summary of the Invention
- Brief Description of the Drawings (if filed)
- Detailed Description
- Claim(s)
- Abstract of the Disclosure

4. ☐ Drawing(s) (35 U.S.C. 113) [Total Sheets ____]

5. Oath or Declaration [Total Pages ____]
a. ☐ Newly executed (original or copy)
b. ☐ Copy from a prior application (37 CFR 1.63 (d)) (for continuation/divisional with Box 18 completed)
    i. ☐ **DELETION OF INVENTOR(S)**
    Signed statement attached deleting inventor(s) named in the prior application, see 37 CFR 1.63(d)(2) and 1.33(b).

6. ☐ Application Data Sheet. See 37 CFR 1.76

ADDRESS TO:
Assistant Commissioner for Patents
Box Patent Application
Washington, DC 20231

7. ☐ CD-ROM or CD-R in duplicate, large table or Computer Program (Appendix)

8. Nucleotide and/or Amino Acid Sequence Submission (if applicable, all necessary)
a. ☐ Computer Readable Form (CRF)
b. Specification Sequence Listing on:
   i. ☐ CD-ROM or CD-R (2 copies); or
   ii. ☐ paper
c. ☐ Statements verifying identity of above copies

ACCOMPANYING APPLICATION PARTS

9. ☐ Assignment Papers (cover sheet & document(s))
10. ☐ 37 CFR 3.73(b) Statement (when there is an assignee)   ☐ Power of Attorney
11. ☐ English Translation Document (if applicable)
12. ☐ Information Disclosure Statement (IDS)/PTO-1449   ☐ Copies of IDS Citations
13. ☐ Preliminary Amendment
14. ☐ Return Receipt Postcard (MPEP 503) (Should be specifically itemized)
15. ☐ Certified Copy of Priority Document(s) (if foreign priority is claimed)
16. ☐ Nonpublication Request under 35 U.S.C. 122 (b)(2)(B)(i). Applicant must attach form PTO/SB/35 or its equivalent.
17. ☐ Other: ....

18. If a CONTINUING APPLICATION, check appropriate box, and supply the requisite information below and in a preliminary amendment, or in an Application Data Sheet under 37 CFR 1.76:
☐ Continuation ☐ Divisional ☐ Continuation-in-part (CIP) of prior application No.: ___/___

Prior application information: Examiner ____ Group Art Unit ____

For CONTINUATION OR DIVISIONAL APPS only: The entire disclosure of the prior application, from which an oath or declaration is supplied under Box 5b, is considered a part of the disclosure of the accompanying continuation or divisional application and is hereby incorporated by reference. The incorporation can only be relied upon when a portion has been inadvertently omitted from the submitted application parts.

19. CORRESPONDENCE ADDRESS

☐ Customer Number or Bar Code Label (Insert Customer No. or Attach bar code label here)   or ☐ Correspondence address below

Name		
Address		
City	State	Zip Code
Country	Telephone	Fax
Name (Print/Type)	Registration No. (Attorney/Agent)	
Signature	Date	

Burden Hour Statement: This form is estimated to take 0.2 hours to complete. Time will vary depending upon the needs of the individual case. Any comments on the amount of time you are required to complete this form should be sent to the Chief Information Officer, U.S. Patent and Trademark Office, Washington, DC 20231. DO NOT SEND FEES OR COMPLETED FORMS TO THIS ADDRESS. SEND TO: Assistant Commissioner for Patents, Box Patent Application, Washington, DC 20231.

Sample Form (former PTO/SB/41) (05-01)

ASSIGNMENT OF PATENT

Docket Number (optional)

Whereas, I, _____ of _____, hereinafter
referred to as patentee, did obtain a United Stated Patent for an improvement in _____
_____

No._____ , dated _____ ; and whereas, I am now the sole
owner of said patent, and,

Whereas,_____

of _____

hereinafter referred to as "assignee" whose mailing address is

_____

City of _____, and State of _____
is desirous of acquiring the entire right, title and interest in the same;

Now, therefore, in consideration of the sum of _____dollars ($_____), the receipt whereof is
acknowledged, and other good and valuable consideration, I, the patentee, by these presents do sell, assign and transfer
unto said assignee the entire right, title and interest in and to the said Patent aforesaid; the same to be held and enjoyed
by the said assignee for his own use and behoof, and for his legal representatives and assigns, to the full end of the term
for which said Patent is granted, as fully and entirely as the same would have been held by me had this assignment and
sale not been made.

Executed this _____day of _____, 20 _____.

at _____.

_____
(Signature)

State of_____)
County of_____)   SS:
Before me personally appeared said_____
and acknowledge the foregoing instrument to be his free act and deed this _____day
of _____.20 _____.

_____
Seal   (Notary Public)

This form offers a sample or suggested format for an assignment document. This sample form is not an OMB officially approved form.

About the author

Jeffrey H. Matsuura is an assistant professor of law and the director of the program in law and technology at the University of Dayton School of Law. Mr. Matsuura is also Of Counsel to the technology law firm the Alliance Law Group, in Vienna, Virginia, where he specializes in providing legal and business counsel to enterprises involved in the information technology, telecommunications, and digital media industries. He is the author of *Security, Rights, and Liabilities in E-commerce* and *A Manager's Guide to the Law and Economics of Data Networks*, both published by Artech House. He is also a coauthor of *Law of the Internet*.

Mr. Matsuura has worked as counsel for several technology-based companies, including Satellite Business Systems, MCI Communications Corporation, Discovery Communications, COMSAT, and TELE-TV. Mr. Matsuura has served as an adviser to the Virginia legislature's Joint Commission on Technology and Science and to the National Task Force on Knowledge and Intellectual Property Management. He has earned a B.A. from Duke University, a J.D. from the University of Virginia, and an M.B.A. from the Wharton School at the University of Pennsylvania. Mr. Matsuura can be reached at jmatsuura@alliancelawgroup.com.

Index

Technology export controls (continued)
  Wassenaar Arrangement and, 149–50
  See also Export controls
Technology taxes/fees, 161
Tivo, 99
Top-level domains (TLDs), 62–63
  country code, 64–66
  defined, 62
  generic, 64
  simulated, 62
  See also Domain names
Trade dress, 5
  in look/feel context, 82–83
  protection, 82
  rights, 82
Trademark dilution, 5
  claim, 52
  trademark owner remedy, 52
Trademark law, 5, 49–70
  defined, 50–51
  domain names and, 55
  importance, 49
  online keyword searches and, 67
  protection, 50–51, 81–83
  rights, 50
  for specific elements of online material, 81
Trademarks
  cybersquatting and, 56–59
  defined, 49
  distinctive, 52
  domain name registration process and, 54–55
  forms, 49–50
  generic meanings, 51
  geographic extent, 53
  hypertext links and, 66
  keywords and, 67
  misappropriation, 56
  multiple owners, 63
  overview, 49–53
  owner proof of, 51
  registration, 53
  rights, 5–6
  terms, 53

Trade secrets
  disadvantages, 22
  law, 6–7
  misuse, 21
  protection difficulty, 22
  protection for computer code, 20–23
  rights, exercise of, 23
  security, 20
  term of, 22
  theft penalties, 21
  types of, 20
Trespass to property, 84–85

U
Unfair competition claims, 68
Unfair competition law, 68
Uniform Computer Information
        Transaction Act (UCITA),
        139–41
  adoption debate, 141
  copyright law and, 140, 141
  defined, 140
  legal rights establishment, 140
Uniform Dispute Resolution Procedures
        (UDRP), 59–61
  arbitration process, 60
  cybersquatting remedies, 59
  defined, 59
U.S.
  business method patents, 18
  copyright law, 12–15
  Copyright Office, 97–98
  moral rights, 23
  patent law, 16
User interfaces
  business method patents and, 77, 79
  design patents and, 80
  See also Look/feel
U.S. Patent and Trademark Office, 20
Utility patents, 79

V
Valuation, 167–71, 191
Valuation models, 167–71
  asset, 170

Artech House Communications Law and Policy Series

Global Telecommunications Market Access, Jennifer A. Manner

International Telecommunications Handbook, Robert M. Frieden

An Introduction to U.S. Telecommunications Law, Second Edition, Charles H. Kennedy

The Law and Regulation of Telecommunications Carriers, Henk J. Brands and Evan T. Leo

A Manager's Guide to the Law and Economics of Data Networks, Jeffrey H. Matsuura

Managing Intellectual Assets in the Digital Age, Jeffrey H. Matsuura

Managing Internet-Driven Change in International Telecommunications, Rob Frieden

Practical Internet Law for Business, Kurt M. Saunders

Securities, Rights, and Liabilities in E-Commerce, Jeffrey H. Matsuura

Telecommunications Deregulation and the Information Ecomomy, Second Edition, James K. Shaw

For further information on these and other Artech House titles, including previously considered out-of-print books now available through our In-Print-Forever® (IPF®) program, contact:

Artech House
685 Canton Street
Norwood, MA 02062
Phone: 781-769-9750
Fax: 781-769-6334
e-mail: artech@artechhouse.com

Artech House
46 Gillingham Street
London SW1V 1AH UK
Phone: +44 (0)20 7596-8750
Fax: +44 (0)20 7630-0166
e-mail: artech-uk@artechhouse.com

Find us on the World Wide Web at:
www.artechhouse.com